Critical Theory and the Thought of Andrew Feenberg

Darrell P. Arnold · Andreas Michel
Editors

Critical Theory and the Thought of Andrew Feenberg

palgrave
macmillan

Editors
Darrell P. Arnold
Biscayne College
St. Thomas University
Miami Gardens, FL, USA

Andreas Michel
Rose-Hulman Institute of Technology
Terre Haute, IN, USA

ISBN 978-3-319-57896-5 ISBN 978-3-319-57897-2 (eBook)
DOI 10.1007/978-3-319-57897-2

Library of Congress Control Number: 2017944096

Cover design by Samantha Johnson

Printed on acid-free paper

This Palgrave Macmillan imprint is published by Springer Nature
The registered company is Springer International Publishing AG
The registered company address is: Gewerbestrasse 11, 6330 Cham, Switzerland

Contents

1 **Introduction** 1
Darrell P. Arnold and Andreas Michel

Part I Reverberations of Marx and Lukács

2 **What is the "Philosophy of Praxis"?** 17
Steven Vogel

3 **Philosophy of Praxis or Philosophical Anthropology? Andrew Feenberg and Axel Honneth on Lukács's Theory of Reification** 47
Konstantinos Kavoulakos

4 **Gegenständlichkeit—From Marx to Lukács and Back Again** 71
Christian Lotz

5 **Feenberg, Rationality and Isolation** 91
Clive Lawson

Part II Between Democracy and a Politics of Resistance

6 Transforming Dystopia with Democracy: The Technical Code and the Critical Theory of Technology 117
Graeme Kirkpatrick

7 Andrew Feenberg's Ecological Modernism 139
Darrell P. Arnold

8 Between Instrumentalism and Determinism: Western Marxism and Feenberg's Critical Theory of Technology 163
Matthew Greaves

Part III From Heidegger Through Postmodernism: The Ongoing Role of the Critical Theory of Technology

9 The Question Concerning a Vital Technology: Heideggerian Influences on the Philosophy of Andrew Feenberg 193
Dana S. Belu

10 Future Questions: Democracy and the New Converging Technologies 217
Andreas Michel

11 Revisiting Critical Theory in the Twenty-First Century 241
Raphael Sassower

12 Andrew Feenberg, Critical Theory, and the Critique of Technology 263
Douglas Kellner

Part IV A Critical Response

13 Replies to Critics: Epistemology, Ontology, Methodology 285
Andrew Feenberg

14 **Appendix: Interview with Bruna Della Torre de Carvalho Lima and Eduardo Altheman Camargo Santos** 319
Bruna Della Torre de Carvalho Lima and Eduardo Altheman Camargo Santos

Index 327

Editors and Contributors

About the Editors

Darrell P. Arnold is Editor of *Traditions of Systems Theory* (Taylor & Francis) and translator of many works from German, including Matthias Vogel's *Media of Reason* (Columbia UP) and Chrys Mantzavinos' *Naturalistic Hermeneutics* (Cambridge UP). He works on issues in critical theory and systems theory, especially as relevant to globalization, environmental philosophy, and philosophy of technology. Darrell is Professor of Philosophy and Coordinator of the Liberal Studies and Philosophy programs at St. Thomas University. He is also the President of the Humanities and Technology Association (HTA), a national interdisciplinary society focusing on the study of technology and society. From 2012 through 2014 Darrell was editor of the *Humanities and Technology Review*, the journal of the HTA.

Andreas Michel is Professor of German at Rose-Hulman Institute of Technology in Terre Haute, Indiana. His main interests are intellectual history, theories of modernity, and the philosophy of technology. He has recently published *Historiografie der Moderne: Carl Einstein, Paul Klee, Robert Walser* (eds. Michael Baumgartner, Andreas Michel, Reto Sorg) Fink Verlag, 2016. He has published essays on Carl Einstein, Gianni Vattimo, German romanticism, modernity and postmodernity, and the philosophy of technology. He has served as President of the Humanities and Technology Association (HTA) (2006–2012) and of the Carl Einstein-Society (2010–2017).

Contributors

Dana S. Belu is Associate Professor of Philosophy in the Philosophy Department at *California State University, Dominguez Hills*. She works at the intersection of continental philosophy, philosophy of technology and feminist philosophy. Her book *Heidegger, Reproductive Technology & The Motherless Age* was published with Palgrave Macmillan in 2016.

Andrew Feenberg is Canada Research Chair in Philosophy of Technology in the School of Communication, Simon Fraser University, where he directs the Applied Communication and Technology Lab. He also serves as Directeur de Progamme at the Collège International de Philosophie in Paris. His books include *Questioning Technology, Transforming Technology, Heidegger and Marcuse, Between Reason and Experience*, and *The Philosophy of Praxis: Marx, Lukács and the Frankfurt School*. A book on Feenberg's philosophy of technology entitled *Democratizing Technology*, appeared in 2006. His book *Technosystem: The Social Life of Reason* was published in 2017.

Matthew Greaves is a doctoral candidate, instructor, and union activist at Simon Fraser University in Vancouver, Canada. His research appears in journals such as *Rethinking Marxism, The International Journal of Technology, Knowledge and Society, New Proposals, and the Canadian Journal of Communication.*

Konstantinos Kavoulakos (born in Athens, Greece, in 1967) is Associate Professor of Social and Political Philosophy at the University of Crete (Greece). He is the author or editor of 12 books. His research focuses on an investigation of the twentieth century tradition of critical social thought. His most recent books include *Ästhetizistische Kulturkritik und ethische Utopie. Georg Lukács's neukantianisches Frühwerk*, (Berlin/Boston: de Gruyter, 2014), and *Tragedy and History. The Critique of Modern Culture in the Early Work of Georg Lukács 1902–1918* (Athens: Alexandria, 2012, in Greek).

Douglas Kellner is George Kneller Chair in the Philosophy of Education at UCLA and is author of many books on social theory, politics, history, and culture, including *Herbert Marcuse and the Crisis of Marxism, Camera Politica; Critical Theory, Marxism, and Modernity; Jean Baudrillard: From Marxism to Postmodernism and Beyond*; works in social theory and cultural studies such as Media Culture; a trilogy of books on postmodern theory with Steve Best; a trilogy of books on the media and the Bush administration; his most recent books are *Media Spectacle and Insurrection, 2011: From the Arab Uprisings to Occupy Everywhere and American Nightmare: Donald Trump, Media Spectacle, and Authoritarian Populism*. His website is http://www.gseis.ucla.edu/faculty/kellner/kellner.html.

Graeme Kirkpatrick is senior lecturer in Sociology at the University of Manchester. He is author of several monographs on critical theory, aesthetics and technology, including *Aesthetic Theory and the Video Game* (MUP 2011) and *Computer Games and the Social Imaginary* (Polity 2013).

Clive Lawson is the Director of Studies in Economics at Girton College, Cambridge, UK. His central interest is in social ontology, but he has written recently in environmental economics, the philosophy of technology and the history of economic thought. He is an editor of the Cambridge Journal of Economics and founder member of the Cambridge Social Ontology Group. His book *Technology* and *Isolation* was published with Cambridge University Press in 2017.

Christian Lotz is Professor of Philosophy at Michigan State University. His most recent book publications are *The Art of Gerhard Richter. Hermeneutics, Images, Meaning* (Bloomsbury Press, 2015, pbk 2017); *The Capitalist Schema. Time, Money, and the Culture of Abstraction* (Lexington Books, 2014; pbk. 2016); *Christian Lotz zu Marx, Das Maschinenfragment* (Laika Verlag, 2014); *Ding und Verdinglichung. Technik- und Sozialphilosophie nach Heidegger und der kritischen Theorie* (ed., Fink Verlag, 2012). His current research interests are in classical German phenomenology, critical theory, Marx, Marxism, aesthetics, and contemporary European political philosophy.

Raphael Sassower is Professor and occasional Chair of Philosophy, and Director of the Center for Legal Studies at the University of Colorado, Colorado Springs. Concerned with postmodern technoscience, the range of topics he writes about include technoscience, political economy, aesthetics, and education. In addition to writing journalistic pieces, his latest books include: *Democratic Problem-Solving: Dialogues in Social Epistemology* with Justin Cruickshank (2016), *Compromising the Ideals of Science* (2015), and *The Price of Public Intellectuals* (2014).

Steven Vogel is John and Christine Warner Professor of Philosophy at Denison University. He is the author of *Thinking like a Mall: Environmental Philosophy after the End of Nature* (MIT Press, 2015) and *Against Nature: The Concept of Nature in Critical Theory* (SUNY Press, 1996), as well as of many articles in journals such as *Environmental Ethics, Environmental Values*, and others. He is also the Co-Director of the International Association for Environmental Philosophy.

CHAPTER 1

Introduction

Darrell P. Arnold and Andreas Michel

Andrew Feenberg has long made a valuable contribution to critical theory. His particular focus has been on the issue of technology that original members of the Frankfurt School viewed as vitally important, but that have seldom been given systemic treatment in later largely Habermas-inspired versions of critical theory. Feenberg, who studied with Marcuse, has not only followed in the footsteps of the original Frankfurt School in devoting considerable systemic thought to this issue, but along with Steven Vogel and Douglas Kellner, who both contribute to this volume, he has in other ways further developed a form of critical theory more Marcusean than Habermasian in inspiration. More than merely developing critical theory of technology, Feenberg has also been an astute interpreter of major critical theorists—or philosophers of praxis—including Marx, Lukács, and the Frankfurt School.

Though one volume has been published on Feenberg's contribution to philosophy, that work is focused on Feenberg's philosophy of

D.P. Arnold (✉)
St. Thomas University, Miami Gardens, FL, USA

A. Michel
Rose-Hulman Institute of Technology, Terre Haute, IN, USA

D.P. Arnold and A. Michel (eds.), *Critical Theory and the Thought of Andrew Feenberg*, DOI 10.1007/978-3-319-57897-2_1

technology. The idea for the present volume is different. Various essays here do engage Feenberg's views on technology. However, in general, the volume casts a broader net. This book was conceived after Feenberg's publication of *Philosophy of Praxis: Marx, Lukács, and the Frankfurt School*, a volume in which Feenberg particularly displays his skills as a philosophical interpreter and theoretician within the philosophy of praxis. Many of the contributions here take up those broader issues of critical theory, even when they also dovetail into questions of technology. Part I of this volume contains essays focused on Feenberg's readings of Marx and Lukács, and the continued relevance of these views. Part II focuses on Feenberg's explicit views on democracy. Part III examines Feenberg's views particularly in light of contemporary developments in Continental phenomenology and postmodernism and in reference to pertinent discussions in philosophy of technology.

Part I: Reverberations of Marx and Lukács

In the opening essay for the volume, "What is the Philosophy of Praxis?" Steven Vogel sets out to discuss in explicit terms what philosophy of praxis is. He highlights that despite the title of Feenberg's 2014 monograph, *The Philosophy of Praxis: Marx, Lukács and the Frankfurt School*, Feenberg nowhere in that book clearly defines the term "praxis." After some introductory comments about the importance of Feenberg's work and influence, Vogel takes it upon himself to describe the term praxis, then to explain kernel ideas within critical theory based on that idea. As Vogel notes, the term is used to describe Marx's action-oriented understanding of human life: "humans are constantly acting, constantly *doing things*, and 'practice' is the name this theory gives to those doings" (p. 20). Vogel highlights the "transformative" character of such activity in the world and goes on to describe Marx's epistemology and ontology as well as that of many writings in the philosophy of praxis. In Vogel's view, though Feenberg's contribution to understanding the philosophy of praxis is considerable, he is incorrect to identify Marx's view of "labor" too narrowly with wage labor. Vogel suggests that we rather interpret Marx's use of the term "labor" as equivalent to "praxis." Marx, in Vogel's view, sees us as laboring in the world in various domains—that is, he views us as engaged in practices in the world through which we transform the world. Vogel clearly narrates the story of epistemology from British empiricism through Kantian and post-Kantian idealism to

Marx's and Marx-inspired materialism. He highlights how the Marxist tradition comes to see German idealism and early modern epistemology as fostering merely "contemplative" forms of life. In laying out this narrative, Vogel offers a clear discussion of Marx's view of alienation, showing its similarities to Lukács's treatment of reification. While the majority of Vogel's account is sympathetic to Feenberg's work, with various points of difference highlighted, in the final two sections of his essay, Vogel takes particular issue with Feenberg's understanding of nature, arguing, in contrast to Feenberg, that we are incorrect to view the natural world as distinct from the social world. Vogel argues, "there is only one world, not two" (p. 35). While we obviously do not create a world ex nihilo, our labor and historically informed mediation of reality are fundamental to the one world that is intricate(ly) tied to human practices. This essay provides an excellent overview of the discussion of praxis and an intricate(ly) analysis of the relationship between nature and culture as it plays out in the work of Marx, Lukács, Feenberg, and Vogel.

In "Philosophy of Praxis or Philosophical Anthropology," Konstantinos Kavoulakos examines the resurgence of interest in Lukács's view of reification, in part inspired by Feenberg's own treatment of the concept. His essay compares Feenberg's and Axel Honneth's treatments of reification and argues that Feenberg's interpretation offers a less ideological understanding of the concept than Honneth's and a more helpful one in light of the present needs to critically transform global capitalism. In Kavoulakos' view, it is of particular interest that Feenberg emphasizes that a final dereified worldview, brought on, for example, by a great revolutionary struggle, is never possible, but rather that dereifying our worldviews through a process of political struggle is a constant and never-ending process. Marx's promised proletarian revolution never occurred and is hardly expected. What we are left with in its place are fragmented political struggles. In such a context, rather than expecting revolutionary action that will lead to the final overcoming of a reified worldview, we might instead expect "transformative" action in multiple domains, creating "a framework of meaning within which activity goes on" (see here, p. 53, fn. 34).

Kavoulakos contrasts Feenberg's dialectical and praxis-oriented reading of reification with Honneth's reading, arguing that Honneth abandons the dialectic holistic impulse of radical critical thought altogether and comes to view reification largely as designating "a cognitive occurrence in which something that doesn't possess thing-like characteristics in itself (for example, something human) comes to be regarded as a thing" (quoted on p. 54). In Kavoulakos' view, Honneth's interpretation confuses reification with objectification—a kind of negative mental habit that prohibits an "emphatic engagement" with people and objects. Kavoulakos argues that this in part derives from Honneth's failure to appreciate the fundamental place in Lukács's philosophy of the connection between social institutions, cultural structures, and human behavior. Honneth's perspective ultimately slips into anthropology, which Lukács himself saw as dangerous precisely because anthropology always tends to become reifying. Kavoulakos contrasts what he views as the non-dialectical position of Honneth with Feenberg's view, arguing that Feenberg's retention of a dialectical perspective provides a theoretical tool for confronting the irrationality of not only the technical system that Feenberg emphasizes but by extension the economy as well. Such rational systems are only partially differentiated from the lifeworld. They still retain the potential for "transformation from below" (p. 63). While Kavoulakos is very sympathetic to Feenberg's views of reification, he emphasizes the importance of applying Feenberg's ideas to the economy more explicitly than Feenberg himself does.

In "Gegenständlichkeit—From Marx to Lukács and Back Again," Chap. 4 of the volume, Christian Lotz argues that Marx and Lukács should both be understood as engaged in social ontology. He maintains that Feenberg rightly points out that the "form of objectivity" in Lukács is the form through which we understand and deal with objects in the world. Nonetheless, he thinks Feenberg would benefit from focusing more attention on Lukács's early writings in *The History of Class Consciousness*, prior to the reification chapter, and from tying Lukács's views on the form of objectivity back to Marx's writings in *Capital*. It is in *Capital*, Lotz's argues, that Marx develops a social ontology that views existing social structures as categories that serve as the conditions against which the experience of objects is made possible. *Capital* is thus not best read merely as an economics text but as a social ontology that unveils how social structures bring forth objects of particular kinds, understood against the functions that those objects play in the larger social totality. In line with this, concepts are not viewed as changing in isolation but always as part of an interlinked web, with social categories as basic to the production of meaning and action. Lotz argues that while Feenberg provides an important analysis of Lukács, he nonetheless does not offer a reading that clearly enough shows the integral link between social structures, forms of meaning, and the practical uses of objects. These form a totality. In Lotz's view, Feenberg has focused on technology, often at the expense of political economy, a point also underlined by Kavoulakos. Consequently, Feenberg has not given these diverse issues their needed attention.

In Chap. 5, "Critical Theory, Reification, and Isolation," Clive Lawson takes aim at some details of Feenberg's view of social rationality. In particular, he questions the adequacy of Feenberg's views on the nature of science and technology, disputing whether these domains are as strictly delimited as Feenberg maintains. While Lawson does not maintain that there are no important differences between science and technology, he does argue that the domains are more similar than Feenberg acknowledges. Lawson argues that both science and technology entail processes that, in his description of primary and secondary instrumentation, Feenberg describes as applying only to technologies.

In Lawson's view, Feenberg views social rationality as entailing three science-like principles: a "comparison of equivalents; classification and application of rules; [and] the optimisation of effort and calculation" (p. 98). Each of these principles is modeled on science; and science, in this account, is viewed as the benchmark for the judgment of such rational processes. Lawson maintains that Feenberg presumes rather than argues this perspective. But the critical realist conception of science of Roy Bhaskar provides an explicit grounding for these intuitions that, with a slight repositioning, could be aligned with Feenberg's philosophy: "Instead of talking of these [above-noted] principles as rational because they resemble science, Feenberg could argue that his principles resemble the kind of science that can be pursued when *isolatability* is present, and so closed systems are a possibility" (p. 102). Lawson describes the type of science of closed systems he has in mind and argues that closed systems and isolatability are essential for both science and technology. In science, it is the isolation of cases and the use of deduction that allows the formulation of law-like statements. In technologies, Feenberg does account for this type of isolatability at the level of primary instrumentalization. Lawson, however, suggests that just as there is this level of isolatability in science and technology, so there is also a level of what Feenberg calls "secondary instrumentalizaion" not only in technology, as Feenberg notes, but also in science. What Lawson notes as the context for such value ascription of technologies also applies to the meanings of science: "[f]or new devices and mechanisms to be realized, or to *work*, they must be enrolled in particular networks of interdependencies or sets of social relations" (p. 103).

In light of this secondary determination of technologies, Feenberg, in Lawson's view, rightly proposes the democratization of technologies, so that the technologies that affect our lives in such important ways are generated with a view to human needs.

Lawson for his part discusses similarities between science and technology that Feenberg overlooks. Both science and technology have a level of relative freedom from value determinations; and both have a level at which value determinations come into play. While the science of isolatable closed systems has a place, a problem occurs when the methods appropriate for such isolatable systems are applied to observations of social reality where value determinations do in fact play a role. The pretense of such isolatability in part accounts for the reification that occurs in both bureaucracies and in economics. In those areas, it is part of an overall effort to generate power and control over people. In the field of economics (and one might add policy analysis), this attempt to isolate social reality in a system that is not closed in part also accounts for the failure of attempts to adequately describe social reality.

Part II: Between Democracy and a Politics of Resistance

In Chap. 6, "Transforming Dystopia with Democracy: The Technical Code and the Critical Theory of Technology," Graeme Kirkpatrick examines technical codes that exist because of what Feenberg calls the underdetermination of technology. This underdetermination is precisely what allows space for secondary instrumentalization, where values are encoded in technological devices. Because of the underdetermination of primary instrumentalization, values and thus politics always play a role in design. By emphasizing the possibilities for democratic social control over technologies, Feenberg sets himself apart from the dystopian view of most Frankfurt School thinkers. He emphasizes that if such technologies can be brought under democratic control, they can be molded into more humane technological devices. While Kirkpatrick thinks Feenberg clearly does advance beyond the Frankfurt School in emphasizing the more positive opportunity to reshape technologies to meet human needs, he argues that Feenberg still retains too much of the earlier dystopic Frankfurt School view regarding his understanding of technological rationality; for despite Feenberg's position that technological designs are socially encoded, in Kirkpatrick's view he still ontologizes hegemonic technological reason, then contrasts this not quite properly understood view of instrumentalist reason with a non-instrumentalist perspective that can be facilitated by democracy.

Kirkpatrick maintains that in some cases technologies are so underdetermined that mere instrumental control is not an option. While he agrees with Feenberg that the domain of technology design becomes a battlefield between cultural values, he argues that Feenberg does not clearly see the implications of this view. Further, by contrasting hegemonic technical control with democracy, Kirkpatrick thinks Feenberg shows himself overly optimistic that democracy will result in designs that fulfill human potentials. While democratic control may be a value in itself, in fact, democratic participation has not always led to positive results. Nor is there anything to ensure that it will. In addition, democracy alone is insufficient and must be augmented by other means of value determination. In a view that compliments Lawson's, Kirkpatrick argues that Feenberg would benefit from even more clearly articulating the underdetermination even of technological rationality. Further, he would do well to emphasize that democratic procedure alone is insufficient for determining technologies. In many cases more overt political struggle, rather than democratic procedure, will be necessary in order to get the humanizing technologies that are needed.

In Chap. 7, Darrell Arnold begins by laying out Feenberg's optimistic views as to how to make sustainable environmental policies compatible with modern capitalism and modern technology. Far from advocating eco-primitivism or a return to utopian notions of an intact community in harmony with nature, Feenberg's writings put their hopes in democratically administered modern technology that will eventually lead to a paradigm shift in Western societies—away from the focus on efficiency defined in terms of GDP and toward a new understanding of efficiency that involves environmental sustainability: future technological development will support democratically designed environmental policies around post-capitalist values. For his analysis, Feenberg relies on an updated account of Frankfurt School Critical Theory, mainly the writings of Adorno and Marcuse. Along with them, Feenberg places his hopes in the structural and cultural transformation of Western capitalism driven by a respect for nature. In the following passage, Arnold summarizes the main ideas from one of Feenberg's most recent texts, the "Liberation of Nature:" "A renewed approach to nature, informed by aesthetic considerations, would mean that we would be more likely to shape nature in a manner beneficial to its own flourishing and purposes that attune with a broader understanding of well-being (in which having a flourishing environment is viewed as vital to having a good life more generally)" (p. 146).

In the second part of his paper, Arnold takes up some of the "systemic issues" involved in such a new attunement to nature. The first issue is the question as how best to advocate for the ideas of a greening democracy. Here, he contrasts Feenberg's emphasis on deliberative democracy with advocates of administrative rationalism, that is, those who propose that experts should lead us in these decisions without the input of laypeople. With some caveats, Arnold comes out on the side of deliberative democracy while emphasizing that experts should naturally be part of the deliberative process. Second, when theorizing democratic environmental policy, Arnold takes up the question of whether "mankind" ought to function as the supreme value of nature, or whether this place should rather be held by all planetary life. Arnold here contrasts Feenberg's views (which are mostly stated from an anthropocentric position) with the non-anthropological positions of Eckersley and Dryzek, who argue for the broadest possible understanding of nature in environmental deliberation. Arnold then addresses the governmental challenges of environmental policies in a world system of localized nation states. Here, he reminds us of the crucial role such civil society organizations as Greenpeace and Earthfirst! play in the transnational efforts at greening democracy. Though not as hopeful as Feenberg, Arnold nevertheless trusts that the "cultural changes that can be leveraged to help improve our current statist and corporate systems are also the ones that can help create a more global cultural identity that is a prerequisite for an eventual Global Parliament and environmentally sensitive economy" (p. 156).

In Chap. 8, "Between Instrumentalism and Determinism: Western Marxism and Feenberg's Critical Theory of Technology," Matthew Greaves draws out a genealogy of "foreclosure theory," a view that meaningful political action occurs in spaces outside of the system of communication technologies, in the space foreclosed from the marginalized. Here, he places Feenberg's discussion within contemporary Leftist treatments of technology, from those of Antonio Negri, Michael Hardt, and Gilles Deleuze, to Jodi Dean and Christian Fuchs, to Herbert Marcuse and Johan Söderberg. Greaves argues that contemporary Leftist philosophy has much to teach about the entrapments of our technological system(s), but that it also points to possibilities for transforming technology. Here, Feenberg's position offers important theoretical insights that balance the positions of Dean and Fuchs, on the one hand, and the positions of Negri and Hardt, on the other. While Dean and Fuchs falsely exaggerate the extent to which digital technologies are determined to produce alienation, Negri and Hardt find themselves on an opposing side of the continuum, viewing the multitude as a hybrid of the technological and the human, but here seeing the technical as merely instrumentally good or bad. Feenberg offers a path between such determinism and instrumentalism. His thought can thus serve as a basis for an improved disclosure theory that points toward how we can further develop and use technologies, including digital technologies, and do so in ways that may still be useful for class struggle.

Part III: From Heidegger Through Postmodernism: The Ongoing Role of the Critical Theory of Technology

In Chap. 9, Dana S. Belu is looking for a "vital" theory of technology, a practical theory for our times; and she argues that Andrew Feenberg has developed such a theory. In her paper, she teases out Feenberg's debts to Heidegger as well as his departures from him.

Approaching her topic in three stages, Belu begins by honing in on the enduring influence of Heidegger for any theory of technology, especially his essay *The Question Concerning Technology*. Feenberg rejects the notion of *Ge-stell*, which, according to Heidegger, defines the modern way of revealing. Like Feenberg, Belu argues that the explanatory reach of the Heideggerian *Ge-stell* is ultimately limited because the focus on efficiency alone cannot capture the true meaning of technology in modern life. In a second step, Belu discusses Heidegger's student and Feenberg's teacher, Herbert Marcuse. Marcuse's *One-Dimensional Man* offers Feenberg the framework of a social and political ontology for rethinking the technological subject in Western, capitalist societies. Entitled "A Phenomenology of Production: Rethinking Potentiality in the Technical World," this section illustrates Feenberg's original reading of Heidegger against the grain. In this, the most technical part of her essay, Belu shows how Feenberg "uses Heidegger against Heidegger by wresting a phenomenology of production from a Heidegger who attempts to distance himself from a philosophy of making. Moreover, by reading Heidegger through the philosophy of Marcuse's social dialectic, Feenberg draws out the ethically normative implications in Heidegger's critique of instrumentalism" (p. 200). In her third section, "The Lifeworld of Things," Belu shows how Feenberg uses the results of his rearticulation of Heidegger and Marcuse for a vital theory of technology. And she focuses on his instrumentalization theory, which, in her view, represents the "disaggregation of the concept of *Ge-Stell*" as conceived by Heidegger (p. 203). In her estimation, Feenberg's two-tiered theory of instrumentalization captures the "world-disclosing social power of technological devices" much better than does Heidegger's *Ge-stell* (p. 203). It thus represents the heart of Feenberg's social phenomenology of technology.

Toward the end of her contribution, Belu applies Feenberg's instrumentalization theory to her own work. Providing an in-depth quotation from her book *Heidegger, Reproductive Technology & The Motherless Age* (Palgrave Macmillan, 2016), Belu demonstrates how Feenberg's theory of primary and secondary instrumentalization can productively be applied to various forms of advanced reproductive technologies such as in vitro *fertilization* (IVF). With the help of Feenberg's conceptual tools, Belu sets out to "critique the misguided, because overly functionalized and reductive interpretation of this procedure" (p. 209). In this fashion, she provides a strong finish to her paper which, on the whole, is a ringing endorsement of Feenberg's work.

In Chap. 10, Andreas Michel, like Belu, looks at the influence of Heidegger on Feenberg's thought; and, like Belu, he pays tribute to Feenberg's accomplishments in developing a critical theory of technology. Yet the argument of the latter part of Michel's article is that if future developments suggested in present convergent (nano-bio-info-cogno) technologies are in fact realized, then Feenberg's hope to transform technology may not come to pass. The developing human-technological matrix strongly suggests that we in fact have not and will not free ourselves from the Heideggerian *Ge-stell.*

The outside perspective from which Michel approaches Feenberg's work is based on a recent text by Yuval Harari entitled *Homo Deus. A Brief History of Tomorrow.* In this popularized account of the potential impact of recent technological advances, Harari makes the case that, at the beginning of the twenty-first century, mankind finds itself at a crossroads.

Without being an alarmist, Harari argues that with the latest technologies mankind is poised to bring the Anthropocene, the time of the reign of biological man, to an end by empowering human substitutes to rise to prominence. He argues this process has already begun. Smart machines and automation have become part of our lives, and Harari demonstrates how the merger of nanotechnology, biotechnology, informatics, and cognitive science could result in handing over human control to human-made artifacts. With the advent of machine learning and the idea that artificial intelligence can potentially replicate the functioning of organic systems without loss, homo sapiens might have reached a stage where more and more tasks are handed over to artificial systems which, because they can process information so much faster than biological systems, will gain the trust of humanity. As Harari shows in the case of the GPS, such tasks administered by human substitutes may run the gamut from servants to independent decision makers.

The question that emerges from this development is mankind's potential surrendering of control to artificial systems. The point where this theory intersects with Feenberg's work is with the notion of experience. Michel argues that Feenberg's optimistic account of the democratic administration of technology hinges upon the notion of independent human experience in the "life world," where aesthetic, ethical, and political concerns are clearly separable from the utilitarian considerations of technical systems. If, however, technical systems, because of their superior powers of data mining and speed of operation, are asked to provide the "raw materials" on which human decision are based, or if, as in the case of GPS, such systems make the decision, what will happen to the notion of independent, human experience in the process? Any final decision as to what to do in a specific situation might still be made by "us," but the criteria for making the decision have been created by formalized algorithms. At that point, Michel suggests, Heidegger's notion of the *Ge-stell*—which is nothing technical but rather the human desire for complete control over all aspects of life (Being)—seems to regain some purchase. Paradoxically, the desire for complete human control may result in the creation of a new species.

In Chap. 11, "Revisiting Critical Theory in the 21st Century," Raphael Sassower positions Feenberg's critical theory within the tradition of the Frankfurt School, highlighting Feenberg's valuable contribution to that tradition and to critical thought about technology. However, Sassower goes on to argue that Feenberg's strict separation between science and technology is not tenable. Sciences and technologies are far too intricately connected for us to feasibly maintain that they operate according to disparate logics. Technology is as often the mother of science as its child. One will also stumble when trying clearly to differentiate the work of an engineer from a physicist. Science and technology are best understood as combined in technoscience. Sassower also suggests that, in order to remain relevant in the twenty-first century, critical theory would do well to integrate much of the caution about metanarratives characteristic both of Popper's critical rationalism and of postmodernism. However unorthodox it may appear to unite critical theory with Popper and postmodernism, in Sassower's view, relevant critical thinking about technology in our era needs to integrate a concern with the economic system that characterizes critical theory, the piecemeal trial and error methodology of Popper, and the intellectual humility of postmodernism. Further, despite the frequent dismissal of postmodernism as relativist, Sassower emphasizes the ethical intent of much of its theorizing on power, and argues that many of its best practitioners were molded by the same Marxist tradition that inspired the Frankfurt School. Sassower emphasizes that theorizing with a focus on the relationship between technoscience and social and economic power, as emphasized by Feenberg and critical theory more generally, is needed today as much as ever, but that such theorizing has to be context-sensitive, humble, and non-dogmatic. Unlike traditional Marxism, critical thinking about technoscience in the twenty-first century will not likely be able to wield the grand reading of history or capitalism as a dialectical struggle between the oppressive capitalist and an oppressed proletariat that will work toward its emancipation through social struggle. Rather, the power relations have to be rethought in new terms, as does the technoscientific critique aiming at improving the human condition.

In Chap. 12, Douglas Kellner takes issue with an aspect of Feenberg's theory that many in this volume have discussed—namely Feenberg's instrumentalization theory. Like many contributors to this volume, however, Kellner begins by expressing his appreciation for Feenberg's decades-long dedication to formulate a truly critical theory of technology. He applauds Feenberg's stand against essentialisms of all sorts, rejecting in equal measure Heidegger's (and Ellul's) dystopic technological essentialism, on the one hand, and the technophilic essentialism, where technology functions as the harbinger of human salvation, on the other. Kellner also esteems the critical approach Feenberg takes to social constructivism. While Feenberg profits from a more pragmatic approach to technology in his case studies, he also shows himself critical of the limits of this approach, whose descriptive and empiricist dimensions may run the danger of leaving out the big picture of the social and political implications of technology. Lastly, Kellner appreciates Feenberg's general focus, which is to lobby for a practice of democratic participation in technology design. He sees the strength of Feenberg's work in the concrete analysis of technologies and devices that involve the interaction between state and corporate interests, scientists, and technical designers. In this manner, Feenberg invites all of thus to connect social transformation with technical transformation.

Kellner reserves his critical remarks for the last section of his paper. There, he argues, for example, that Feenberg's optimism about democratic participation in technical processes underestimates the power of technology as a force of domination by downplaying the extent to which capital, the state, and dominant institutions construct technologies in their own interests and resist alternative technologies and reconstruction—although, and despite the fact that the entire focus of Feenberg's work is to point out how capitalist imperatives and biases enter into the design, construction, and implementation of technology.

The second, more technical reservation is to be seen in Kellner's claim of a conceptual limitation he sees in Feenberg's attempt to develop an overarching philosophy of technology. These limitations show up when Feenberg tries to outline a more universal, rather than merely modern, theory of technology. In this attempt, Kellner holds, Feenberg "illicitly smuggles in concepts from modernity into a more general philosophical analysis" (p. 277). As an example, Kellner points to Feenberg's two-tiered instrumentalization theory. There are two aspects to Kellner's critique: the first refers to the term instrumentalization as such. Kellner considers the term ill-chosen for a universal theory of technology since it ought to be reserved for the *modern* specimen of technology (a la *Ge-stell*) alone. Kellner suggests that throughout the ages technology has been much more than merely an instrument (which, to his credit, Feenberg, too, has pointed out in his rereading of Heidegger), and that therefore instrumentalization as a term does a disservice to the rich and broad range that Feenberg, in Kellner's words, "wants to capture in concepts like integration (of technology into everyday life), realization (of values and aesthetic qualities), and democratization (of design, uses, reconstruction)" (p. 279). Second, Kellner holds, that the division into primary and secondary levels (of instrumentalization) seems counter-productive from Feenberg's own perspective. Why not collapse both levels and suggest that in the act of technological creation, aesthetic and ethical dimensions are *equiprimordial* with the merely functional and efficient side.

Kellner's last critical comment is about the cocooning power of modern social media and especially about Facebook. This reflection leads him back to his critique of Feenberg's optimistic stance on democratic interventions into technology. More and more people get their news from a particular set of exclusive media which manufacture their own kinds of alternative news. Thus, fake news played an inordinate role in the 2016 federal elections. With reference to this, Kellner ends his paper by stating: "This striking example suggests that technologies can overwhelm, dominate, and transform individual behaviors, ways of assimilating and disseminating news and information, and transform their very modes of communication and social interaction in ways that can be unhealthy and even destructive for individuals and societies" (p. 280).

Feenberg's Response and Interview

In Chap. 13, Andrew Feenberg provides a summary of the main arguments from this volume. He addresses the concerns of his critics, while clarifying his positions. This book concludes with an interview of Feenberg. This too serves to clarify some of Feenberg's positions—on both issues in critical theory and philosophy of technology.

PART I

Reverberations of Marx and Lukács

CHAPTER 2

What is the "Philosophy of Praxis"?

Steven Vogel

I

Andrew Feenberg is one of the most significant contemporary social philosophers. His work and its influence in the philosophy of technology are well-known; he has developed a set of ideas about technology and its relationship to society that go importantly beyond standard debates about the "neutrality" or "complicity" of modern technology with respect to its social consequences, emphasizing the social complexity of technological developments both in terms of their sources and of the surprising ways in which they themselves transform the social environment in which they operate. Nothing like a Luddite, he has provided careful analyses of contemporary technologies that are remarkably sensitive to both their liberatory and their dangerous aspects. His interest in and connections with the worlds of French and Japanese philosophy, in addition, have provided English-speaking readers access to ideas from those worlds and have enriched his work with intriguing cross-cultural investigations of various technologies.

Feenberg was a student of Herbert Marcuse and is perhaps the leading proponent of the continued importance of the latter's work; in that sense, he is also one of the few figures in the contemporary philosophical

S. Vogel (✉)
Denison University, Granville, OH, USA

D.P. Arnold and A. Michel (eds.), *Critical Theory and the Thought of Andrew Feenberg*, DOI 10.1007/978-3-319-57897-2_2

scene who can be said to continue to work in "classical" critical theory. Most work nowadays with roots in the Frankfurt School is strongly Habermasian and typically treats earlier figures such as Adorno or Marcuse with gestures of vague respect before moving off in directions with which they would likely not have had much sympathy. Feenberg, on the other hand, although acknowledging Habermas's significance, wants very much to argue for the validity of the earlier (more radical, and more explicitly Marxist) approach of the first Frankfurt generation and has done so quite impressively.

But there is another figure, from the period just before the Frankfurt School began, who stands in the background of much of Feenberg's work and whom he has done more than almost anyone else in contemporary philosophy to bring back into serious discussion, and that is Georg Lukács. The work of Adorno, Horkheimer, and Marcuse is unthinkable without Lukács, who recognized the deep Hegelian grounding to Marx's thought before the publication of Marx's early manuscripts in the 1930s made that grounding obvious. The entire tradition of Western Marxism stems from Lukács's work, but in many ways Lukács's brilliance and his significance for serious thought about what critical social theory ought to look like (and about how it is related to critical social *practice*) have been sadly overlooked for many decades. Part of Feenberg's importance is as a figure repeatedly returning to Lukács's ideas and trying to restore to them the crucial place in contemporary philosophy that they absolutely deserve.

This was the topic of his first published book, which appeared in 1981 and was entitled *Lukács, Marx, and the Sources of Critical Theory*.[1] I was writing a dissertation on Lukács at the time, and the book was enormously important to me, helping to clarify some of Lukács's key ideas and therefore to develop my own ideas as well. I first met Andy Feenberg a few years later, when he commented (kindly and helpfully) on the first paper I ever gave at a professional conference. And we have been in touch ever since, recognizing in each other, I think, not only kindred spirits, but also clear enough the issues on which we disagree. I was excited when Andrew revised and republished that first book last year, under the new title *The Philosophy of Praxis*, and I read it with the same intellectual excitement I had felt when first reading the original.[2]

"Philosophy of praxis" is the name that Feenberg gives to the basic set of ideas developed by Lukács in his 1923 *History and Class Consciousness*, ideas that Marx's 1844 manuscripts showed to have been central to the latter's early work as well, and also that play a role in the later theorists of the Frankfurt School (especially Adorno and Marcuse). Feenberg's book is a marvelous historical study, offering subtle critical analyses of the twists and turns as the position is worked out in the thought of these four figures. It is less successful, I think, as a clear presentation of the position as a unified whole. My goal in what follows is to try to present my own view of the basic tenets of the "philosophy of praxis" as such; in doing so, it will become clear where Feenberg and I disagree and where I think the source of our disagreements lie. I consider him an intellectual ally and compatriot; on fundamental issues we really agree, and furthermore where we do not agree I always learn from what he has to say. This essay is dedicated to him in friendship, therefore, and in respect.

II

A curious fact about Feenberg's book is that despite its title, he does not say much about what the "philosophy of praxis" actually *is* or why it has the name it does. Gramsci was the first to use the term, but apparently simply as a euphemism for "Marxism" in an attempt to avoid prison censorship. Feenberg's use is more specific, but oddly enough does not mention praxis at all: "the defining trait of philosophy of praxis," he writes, "is the claim that the 'antinomies' of philosophy can only be resolved in history."[3] This seems like a plausible definition but does raise the question of why such a view should be called a philosophy of *praxis*. Given the name, one might have expected the claim to be that the antinomies can only be resolved in praxis, but Feenberg does not say this. Is praxis the same as "history"? What, actually, *is* "praxis"? Strikingly, the term is defined nowhere in the book, nor does it appear in the index.

"Praxis," of course, is not an English word. In English-language sorts of New Left thinking, it has often been used to mean a specially significant sort of "practice," particularly self-conscious, perhaps, or revolutionary. The word is originally Greek and is an important concept in Aristotle. But it is a perfectly ordinary German word, and the thinkers Feenberg is investigating are all German, so it seems appropriate to understand it in its ordinary German meaning: practice. I shall do so in what follows. And for me, the central thesis of the philosophy of praxis, or of practice, is not the one quoted above about the historical resolution of philosophical antinomies, but rather Marx's crucial remark in the eighth *Thesis on Feuerbach* that "*All mysteries which lead theory to mysticism find their rational solution in human practice and in the comprehension of this practice.*"[4] *Practice*, not history, is the key category for the philosophy of practice: What's striking about Feenberg's book is how small a place that category actually plays in his account.

Marx's thesis is a statement not only about philosophy itself (here called "theory") but also implicitly about human beings. Humans are physical, material beings, first of all: the philosophy of practice is an absolutely materialist theory. But their materiality has to be understood as *active*: Humans are constantly acting, constantly *doing things*, and "practice" is the name this theory gives to those doings. To be a living human being is to be active in the world, and to be active in the world means at the same time to *change the world*. All activity is *transformative* activity; the doings or practices that human beings engage in are constantly altering the world around them. To walk is to compress the pathway on which one travels, scattering or crushing items below one's feet; to breathe is to change the ratio of oxygen to carbon dioxide in the air around one; to talk is to send vibrations tumbling through the air toward the ears of one's interlocutor. The claim here is a straightforward one and applies not only to humans but also to all living organisms: *to be alive is to be active*, and activity transforms the environment in which it takes place.

Humans do more than simply transform their immediate environs through walking and breathing, of course; they can also transform the things around them purposefully, in accordance with plans. They consciously build and restructure the world around them through their practices, anticipating the results of those practices and engaging in them specifically to produce those results. It would be a mistake, however, to understand these sorts of practices as involving something *beyond* practice, some "theoretical" moment that is not itself a practice. "Planning" too is itself an activity—we engage in it with pen and paper, or with spreadsheets, or by talking to one another, or sometimes just by trying things and seeing whether they work or not. Thinking, intending, hoping, believing—all of these, at least for a philosophy of practice committed to materialism, involve *action*, practice: they are all *doings* in the material world. As one thinks one continues to breathe, to gaze in some direction or other, perhaps to type on a computer or write on paper or speak to a collaborator. To intend or to believe or to hope all involve performing certain actions, or trying to perform them (which itself involves action). The idea here is not the simplistic behaviorism that identifies various supposedly mental phenomena with "mere" physical behavior, denying such phenomena any felt first-person character at all, but rather it insists that such felt first-person experiences are never independent of real physical activity, of material doings, and could not exist—could not even be imagined to exist—in the absence of such doings. The point follows directly from the materialism: We are physical, material, living organisms, and every such organism is always (and always already) *active*, in motion, doing.

All I have done in the last few paragraphs is to define what "practice" means for the philosophy of practice; such a definition, as I've said, is unfortunately lacking from Feenberg's discussion. But there is more to the philosophy of practice than this definition, and more too than the implicit claim for the priority of practice as a category. There is also the idea, expressed in the eighth *Thesis*, that such a philosophy can help to resolve the "mysteries that lead theory to mysticism," an idea that involves a particular understanding of the history of modern philosophy and of its own place in that history. Feenberg is particularly good at talking about this element of the account; he summarizes it well by describing Lukács as showing that "Marxism [that is, the philosophy of practice] is the veritable *Aufhebung* of classical German philosophy, arising from its inner dynamic on the basis of its results."[5] The philosophy of practice sees the revolutionary movement that Marx and Lukács both support as a key moment in the history of philosophy, and shockingly even sees the proletarian revolution itself as having philosophical significance. The overthrow of capitalism, for them, *is also the solution to a series of philosophical problems.*

The story is a familiar one. The attempt by the British empiricists to provide an epistemological foundation for the increasingly successful new sciences ends in disaster. If all our knowledge comes from sense-experience, and if that experience is understood as the mind's passive reception of information from an external world of objects—so that achieving valid knowledge requires refusing to impose "subjective" ideas upon that information and instead accepting it just as it presents itself—then knowledge of a world independent of humans turns out to be *impossible.* Instead of justifying the claims of science about the world over rationalist attempts to determine its character a priori, empiricism found itself (as in Locke) unable to explain in what sense a world of material substances could be known to exist at all, or (as in Berkeley) driven to deny that anything exists outside of experience, or finally (as in Hume) forced into a potentially corrosive skepticism not only about the real existence of causal relations in the world but also about the existence of the knowing subject who believes in such relations itself.

This is the context in which Kant founds the tradition of "classical German philosophy," arguing that empiricism's error derives from its view of the subject as a *receiver* of information, and of knowledge as requiring the knower to be *passive*. The fundamental structures of the world we experience—space, time, substance, cause—are there because we put them there, because knowledge is a process in which data from outside the subject are actively formed and organized by the subject. The validity of those structures or categories, and our a priori knowledge of those structures or categories, is guaranteed by the fact that we impose them upon that data. The key insight here is really Vico's: We can only know what we *make*. The world we perceive and experience—the world we inhabit, the world investigated by science—is a world *constituted* by the ego.

But Kant retains a theory of "things in themselves" that stand behind the things we experience, which is to say a noumenal world about which nothing can be known. Feenberg gives a particularly good account of the role the problem of the noumenal realm plays in the Kantian system and especially of the various ways Kant tries to overcome it—in the second *Critique* by the idea that the subject of moral action is the noumenal self that underlies the empirical ego, and in the third one by the appeal to aesthetics as a realm in which not only the form of experience but also its content too is "constituted" by the subject.[6] Schiller and Fichte take these ideas further, as Feenberg shows, but it is really in Hegel that something like an adequate resolution is achieved, with the introduction of a dialectical logic according to which the "otherness" of the thing-in-itself is always only a *relative* otherness, the otherness posited by a particular and limited view of knowledge. At each stage in Spirit's development, Hegel tries to show, it finds itself faced with something it cannot grasp, something beyond its ken—and yet then in a dialectical reversal it overcomes that "beyond" and discovers itself in the very otherness it thought it could not reach, producing yet another stage in which a new otherness will soon reveal itself and then be *aufgehoben* in turn. The goal of the process is Absolute Knowledge, the moment in which all otherness is overcome and shown merely to be stages in the development of Spirit: at that point subject and object turn out to be identical, the (relative) otherness of the object turns out always to be the doing of the subject, and Spirit turns out itself to be nothing other than the world.

With this move, the "thing-in-itself" is rejected, assimilated into the world of Spirit as one of its moments. The Kantian dualism of phenomena and noumena is overcome, but the price seems to be a radical idealism in which the real world is revealed to be the product of Sprit, while the material world turns out to be the one whose "otherness" (that is, whose materiality) disappears in the Absolute. One speculative absurdity—a "real" world inaccessible to human knowledge—is exchanged for another one, in which thought appears as the real content of the world while matter appears merely as thought's alienated product.

Marx's role as the "culmination" of classical German philosophy, on this reading, consists in his inversion of Hegel's idealism, preserving the Kantian insight that knowledge has to be active as well as the Hegelian insight that this requires abandoning dualism and the thing-in-itself but understanding this now in a *materialist* way. The "subject" now is the physical human being, not Spirit—and the "activity" that subject engages in, through which she comes not only to know but also to structure the world around her, is now concrete physical activity, not the mysterious and obscure process whereby an ego "constitutes" a world or Spirit "recognizes itself in its other" but rather simply the practical activities through which human beings work on and transform the world. In the *Theses on Feuerbach* Marx calls those activities "practice," as I have above; elsewhere in his work, though, he refers to them as "labor."

Reinterpreted materialistically, the theses of the classical German tradition make a very different kind of sense. We make the world—and so the Vicoesque idea is relevant—not in the sense of some enigmatic constitutive act but rather by *building* it through our labor. Our every act, as I've argued above, transforms the world; the world around us consists of objects that we have built, objects that express who we are and what we find important, and that also help us to live the lives we want to live. Producing that world, engaging in those practices, laboring, is quite simply what it is to be a human being. And the world that those practices create is *our world*, a world that is *not* other than us, *not* unknowable to us, but that is at the same time perfectly real and material. It is a world in which the problem of the thing-in-itself—of something *below* or *behind* the world we experience—simply does not arise, because the world we experience turns out to be the world *we have built*.

The materialism here is an *activist* one, not a naïve naturalism that treats humans as material objects passively pushed or pulled by external (or internal) forces. Humans maintain their agency, but do so as material beings who express that agency through physical actions. This activist materialism is the translation of idealism's account of the subject into the material realm, as Marx explains in the first *Thesis on Feuerbach*, writing that "the chief defect of all hitherto existing materialism—that of Feuerbach included—is that the thing [*Gegenstand*], reality, sensuousness, is conceived only in the form of the *object* [Objekts] or of *contemplation* [Anschauung], but not as *sensuous human activity, practice,* not subjectively. Hence, in contradistinction to materialism, the active side was developed abstractly by idealism—which, of course, does not know real, sensuous activity as such."[7] The world of matter is not simply a world of objects but of subjects too—it is subjectivity *itself* that needs to be interpreted materialistically. Subjects are to be understood as *material* subjects engaging in practices that are materially transformative of the world around them.

The implications for epistemology here are particularly important. We come to know the world, this view asserts, by *acting in it*—by moving around in it, trying things out in it, discovering what works and what does not. "The question whether objective truth can be attributed to human thinking is not a question of theory," writes Marx in the second *Thesis*, "but is a *practical* question. Man must prove the truth, i.e., the reality and power, the this-worldliness of his thinking, in practice. The dispute over the reality or non-reality of thinking which is isolated from practice is a purely *scholastic* question."[8] In fact *there is no thinking which is isolated from practice*: That's why the question is a scholastic one. Like the idealist versions of activist epistemology, this materialist one intends with the notion of practice to overcome a series of dualisms, including the dualism between subject and object itself. Practice is the intertwining of subject and object, after all: in practice a subject transforms the objective world, thereby producing an object in which the subject is expressed. Furthermore, once we drop the idealist assumption that practice occurs as the result of some prior mental act (an intention or thought) by the subject and realize that the mental act is expressed *in* the practice, we can see that practice transforms and produces the subject too—a subject whose objectivity is essential to it. I am what I do, and so my doings help to create not only the objective environment I inhabit but *me* as well. But that environment of course helps to shape and constrain my doings too—and so, crucially, do the other humans with whom I share it. And so together we help to produce the world we inhabit, through our practices, while that world helps to shape us, and those practices, as well.

The environment we inhabit is built through our practices: this crucial idea immediately complicates standard models of the relation between knower and world that take subjectivity and objectivity to be ontologically distinct. Philosophy professors are always using things like chairs as examples when explaining various epistemological views, asking students how they come to know the chair and what that knowledge consists in, but always implicitly treating it as something purely external to the knowing subjects quizzically examining it. And yet what's rarely mentioned is that the chair was *built* by knowing subjects whose practices in fact made the chair what it is. Its builders built its properties into it, transforming through their practices the materials needed to produce it and thereby creating something whose usability (either for sitting or for employing as an epistemological example) was so to speak guaranteed "a priori." I will not say that noticing this resolves all the problems philosophy professors and students like to find in the "object," but it surely complicates them, and complicates in particular the idea that in this case subject and object are separated by a potentially unbridgeable gap: whatever gap there might have been has, as it were, already been bridged.

That the chair had multiple builders, not just one, is crucial here as well. It would be the rare college classroom where a chair was constructed by a single craftsperson devoted to collegiate carpentry; and even if it were the case, still that person's tools, not to speak of all the other objects—trucks, legal documents, etc.—required to place the chair in the classroom were surely produced by others. The philosophy of practice insists that human practices are typically *social* practices. We build things *together*—first of all in that we do so in direct collaboration with others, but second in that our practices are structured by social understandings and norms, and third because the environment we inhabit is *always already* one that humans have built, and so the very objects on which and with which we work (which is to say, the very objective world we live in) is one in which the practices of other humans have already played a role. Whatever one builds, one builds *out of* and *in the context of* and *along with* and *normatively guided by* others.[9] And so our building activities, our practices, are always social ones. *This* is what it means to say that the environment is a "social construct": not that we impose social "meanings" on the world but that what surrounds us is *literally* constructed in our socially organized practices of labor.

My reconstruction of what "the philosophy of practice" might mean is intended, among other things, to suggest that Feenberg's original definition might benefit from a reformulation: it is not in *history* that philosophical problems are solved, on this view, but rather in *practice*. History is the result of practice, that is certainly true: in a way it is simply the history of various forms of practice. But other things are the result of practice too: houses, say, and cities, and governments, and all the kinds of commodities Marx describes in *Capital*. If we are to be materialists, I might suggest, it is the *material objects* that result from practice on which we ought to focus, not an abstraction like "history." In a sense this is my fundamental criticism of Feenberg: that he does not take sufficiently seriously the material (and also mundane) character of practice. If he did so, I think, a number of problems in his account would be resolved.

One of the problems has to do with the idea of "labor." I have been suggesting here that Marx uses it as a name for practice. In particular for him "labor" does not simply mean the painful, dangerous, and exploitative drudgery that he spends so much energy criticizing capitalism for demanding: rather, especially in his early work, key to his critique is that under capitalism the activity through which humans transform the world and express themselves appears as merely a "means for life" that they run from as soon as they have the chance to avoid it. But Feenberg seems unsure on this point, and criticizes Marx for "hover[ing] between hyperbole and absurdity" in the claims he makes for labor.[10] The translation of the mysterious world-constituting activity of the idealist subject into practical human labor, he argues, cannot be complete, because "the imaginable extension of the concept of an object of consciousness is in truth far greater than that of an object of labor."[11] But this is not true, certainly not if we have given up the dualism that believes there to be a world of "consciousness" separate from the material one, and not if we understand "labor" to mean practical activity in the broad sense. Of course I can think of objects that cannot be worked upon by physical means and with physical tools (because of their size or distance from me, or perhaps their "abstract" and intangible character), but when I think of them the thoughts themselves are *acts*, having roles to play in some activity I am engaging in at the time, whether that be trying to understand some aspect of the universe, making children laugh by coming up with imaginary characters, or taking part in a philosophical argument.

Feenberg says that Marx is "not content to confine human creative powers to the narrow domain that mankind actually and potentially transforms in an imaginable labor process, but wants to extend those powers to 'objective reality everywhere,'"[12] but it is not clear why he thinks this is either hyperbolic or absurd, or what parts of "objective reality" are *not* potentially transformable by human practice. The problem becomes clearer when Feenberg complains that Marx's view involves "the reduction of the human relation to nature to labor," noting that "in everyday coping, play, aesthetic appreciation, recognition, and contemplation humans relate to being perhaps just as fundamentally as they do in labor without attempting to remake objects in their own image."[13] But each of these *are themselves forms of practice*, and indeed—despite the way Feenberg talks about them—are practices that transform the world.[14] Even "contemplation" requires the contemplator to be present in the world, to breathe, to be moving: contemplation too is a form of practice. It is true that typically those who engage in these practices are not "attempting to remake objects in their own image," but it is a mistake to suggest that ordinary labor attempts this either. When I build a bookcase, just as when I admire a sunset or play with a basketball, it is the object itself I am concerned with, not some sort of narcissistic self-duplication. Feenberg is operating with a very limited, and negative, sense of what labor is—indicated, for instance, when he writes against what he sees as Marx's extension of labor as implicated in "reality everywhere" that "the universe is not, in principle, mere raw material: the very idea is either absurd or abhorrent."[15] Once labor is understood as practice, it makes little sense to say that it treats the universe as "mere raw material": the objects upon which labor works may well be themselves filled with meaning and value. It is only *capitalist* labor that treats everything it touches—as well as those doing the touching—as "mere raw material." An enlightened and humane set of practices could well involve deep appreciation for the objects with which they deal, and could even leave certain objects relatively unchanged.

III

The most famous of the *Theses on Feuerbach*, of course, is the final one: "The philosophers have only *interpreted* the world in various ways; the point, however, is to *change* it." Rather than simply an inspiring (if vaguely anti-intellectual) slogan calling readers to move beyond armchair philosophizing toward real revolutionary activism, in the context of the philosophy of practice this remark turns out to have a different and striking meaning, according to which revolutionary political activism appears not as an alternative to philosophy but rather as the latter's *telos*. For if the philosophy of practice involves inverting the idealist picture of *Geist* constituting the world into an activist materialist picture of human beings transforming the world through their practices, and thus asserts the thesis of the fundamental priority of practice over theory, then *that thesis itself cannot simply be a matter of theory*—cannot, that is, be yet another theoretical assertion by a thinker *about* the world of practice, but must rather *itself be practical*, and therefore world-transformative. The rejection of idealism's emphasis on theory cannot itself be (merely) a theory, but must rather be something like an Aristotelian practical syllogism, whose consequence is not a theoretical statement but an *act*—a practice. The move to practice (to "changing the world"), is thus *required by the theory*, or rather is its final result. Theory (or "interpretation" of the world), taken to its logical conclusion, has to transcend theory and become practice.

But what sort of practice? Here the question becomes complicated, not least because the philosophy of practice seems also to assert that all theory is actually a form of (or grounded in) practice, an assertion that renders the distinction between interpreting the world and changing it a bit fuzzy. If all interpretations of the world are based in practice, then they all change the world; what does Marx mean by calling for a new kind of practice that goes beyond mere interpretation? And what happened to the *critical* element of Marx's theory, which after all is the most important part of it—the idea that certain practices, like the ones characteristic of capitalism and commodity production, are unjust and exploitative, and for that matter also the idea that certain interpretations of the world, like the ones that are used to justify that unjust and exploitative social order, are simply *wrong*?

The Hegelian story of the relation of Spirit to "objectivity" involves the repeated discovery by the subject that that which it took to be other than it turns out not to be other at all: in recognizing itself in its other, it moves to a new dialectical stage. In the materialist translation of this story, accordingly, world-transformative practices that know themselves as such, practices in which the subject recognizes and takes responsibility for the changes that are produced, represent higher or better forms of practice than those that are ignorant of what they are and what they do. The normative assumption here dates back to the Delphic oracle: self-knowledge and self-understanding are to be preferred over ignorance and self-deception. This suggests an answer from within the philosophy of practice to the familiar question of what the normative basis is for the Marxist critique of capitalism.

The claim of the philosophy of practice is that humans are fundamentally practical (social) actors in the world and the world is the product of our (socially organized) doings. But there are some doings, it turns out, that do not seem like doings at all, and appear instead in the form of Things, separate from us and independent of our doings. Lukács gives such doings a wonderful name: reification, *Verdinglichung*, "thingification." Marx uses a different name, alienation—but the phenomenon he is describing is the same. The central characteristic of capitalism, according to the Marx of the early manuscripts, is that the objects that humans build through their labor appear to them as external and independent powers—that the more workers transform the world around them the more that world seems like something separate from them and out of their control, indeed as the source of their misery. Their product is not recognizable as such, instead appearing as the wealth of the capitalist that keeps them in submission. This idea is developed further in the mature economic theory of *Capital*, where it is called the fetishism of commodities: the exchange value of a commodity appears as a quasi-natural property of it, instead of an indication of the human labor that produced it. Thus as Marx famously says, a "definite social relation between men … assumes … the fantastic form of a relation between things."[16] The labor by which various human beings collaborated to produce an object that is useful to other human beings appears in the form of a Thing and so is "thingified." And the point of *Capital* is to show that the entire system of market exchange and wage-labor, and with it the oppression and immiseration of the proletariat, follows from this process. It appears as the "natural" form of social organization (subject, for example, to the "Iron Law of Wages"), but in fact is the product of the ways in which human practices are engaged in capitalist society.

Lukács extends Marx's idea, showing how reification characterizes capitalist society at all levels. To be a member of a capitalist society is to confront a series of institutions each of which appears as something like a "fact of nature"—something that is simply given, to whose rules of operation one must adjust oneself, and the question of whose potential mutability or justifiability never arises—despite the fact that those institutions are in fact simply the product of a myriad of social interactions among community members. This is true of that crucial institution called the "market" above all: think of the way we experience "facts" like the rate of growth of GDP, or the inflation or unemployment rate, or for that matter the Dow Jones Industrial Average. These appear as Things that determine our own behavior, that we have no way as individuals to affect but must simply observe and react to, but that in fact of course are themselves simply the aggregated *result* of all of us as individuals observing and reacting in this way, without recognizing that *we ourselves are the authors* of the "alien" phenomenon we face. The individual subject's attitude to the objective world she inhabits—a world she helps to produce but that seems alien, like a piece of nature—is what Lukács calls a "contemplative" one: it involves treating objectivity as something one can only passively respond to but can never actively transform. This is precisely the epistemological position of empiricism, it is worth noticing—the one that was overcome (but only in theory) by the Kantian view of knowledge as active. In the material world it characterizes a set of practices that treat objectivity as something independent of the subject, failing to notice that the world of objects is one that is *socially constructed* in the sense described above.

I have argued elsewhere that environmental problems have their origin in this sort of phenomenon.[17] The basic structure of all these cases is the one that Garrett Hardin named the "tragedy of the commons," where in the absence of a procedure for communal decision-making individuals are faced with a situation in which the act that it is rational for each to perform when aggregated results in an outcome harmful and undesirable to all. I cannot affect global warming by privately deciding not to burn fossil fuels; all that decision does is to worsen my situation significantly while decreasing carbon emissions by an infinitesimal and imperceptible amount. If we agreed to do this together global warming would be decreased, but since there is no way for us to make that communal decision, it is irrational for me to cease burning them—although the consequence of everyone reasoning this way is precisely what produces global warming. The same is true for the factory owner concerned about pollution or the fisherman desiring to preserve fish stocks: Private individuals are faced by a totality that makes it impossible for their private acts to make a difference. The aggregated act of multiple individuals harms everyone, but to each individual that aggregated act appears as a Thing to which she can only respond, and her response (multiplied across all individuals) produces exactly the harmful aggregate itself. The only solution would be for the individuals to decide *as a community* what they wished to do.

This is, in a way, the fundamental argument for socialism: Reification can only be overcome by a communal decision, in which the community as a whole chooses to act self-consciously *as* a community. In the terms of the philosophy of practice this point can be put as follows: The world we inhabit is the product of our socially organized practices. But when we engage in those practices without recognizing this fact, when we fail to acknowledge that the phenomena that surround us (commodities, economic structures, social institutions, changes to the climate) are the products of our own practices, then those phenomena come to look like independent Things with power over and against us that we have no power to question or to change. In accordance with the basic insight of the philosophy of practice, however, this "failure to acknowledge" the world as the product of our practices is *itself a kind of practice*—the kind associated with capitalist free markets, in which private individuals engage in private transactions with each other oriented toward private gain, and no significant space for communal decision-making is to be found. To engage in world-transforming practices self-consciously, on the other hand, would be to engage in them *as a community*, recognizing that *we* (not I) are responsible for the world we inhabit, and to make decisions about those practices democratically and to see them as *our* (not my) practices. No longer appearing as an alien Thing, the practices would appear instead as our own self-expression—as practices that knew themselves to be world-transforming and therefore knew the world we inhabit to be *our* world and not something alien to and beyond us.

When Marx distinguishes between interpreting the world and changing it, we can now see, he is distinguishing not so much between theory and practice as between *two kinds of practice*—the contemplative kind that views the world as independent of our activity and the self-conscious kind that recognizes the product of its activity as its own, and in this sense overcomes reification. I began this section by asking what the practice that Marx is calling for might consist in, and now we have an answer: A set of communal practices that *know themselves to be world-constructing* and that are thus chosen self-consciously by the community through a democratic process of communal decision-making. To say this of course is still to provide no detail about what specific set of practices these might be—but that is not a weakness so much as precisely what the philosophy of practice would lead one to expect. No theoretical analysis prior to the actual communal decision-making itself could provide the answer, because if it could that theoretical answer would have priority over the practical process of deciding, and furthermore would appear to that process as an independent Thing to which the process's relation could only be a contemplative one. Only the community itself can decide what its own practices should be—that decision is a matter of practical democratic choice, not of philosophical argumentation.

The communally self-conscious decision I've been discussing is identified by Lukács with the proletarian revolution itself, and Feenberg does an excellent job of explaining what this means.[18] Among other things as we saw earlier, it ascribes to the revolution a distinctive epistemological and even metaphysical significance, since it would now not merely transform the structure of society or the economy, but is itself also the *solution* to a series of philosophical problems (about form versus content, is versus ought, the relation of subject and object, the nature of the thing-in-itself, etc.). Such a claim about the proletarian revolution, Feenberg recognizes, is terribly hard to believe today, not least because it is no longer clear that the concept "proletarian revolution" has much meaning nowadays at all. He offers an interesting argument to the effect that for Marcuse the formation and growth of the counterculture of the 1960s offered a possible substitute for what Lukács was imagining—but of course this too appears nowadays as at best another failed opportunity.[19] The ultimate paradox of the philosophy of practice is that unless and until the self-conscious communal practices it calls for actually take place all it has to offer is yet another interpretation of the world, and hence by its own standards until that time it remains unjustified and even in a quite serious sense untrue. I have no solution to offer to this paradox, which might indeed be an ultimately tragic one, but will simply leave it stated in this form here.

IV

The key issue on which Feenberg and I disagree has to do with the status of "nature" in the philosophy of practice. To emphasize the idea that human beings *construct the world* raises the question of how literally this is to be understood: Could *nature* be a human construction? Feenberg understandably finds this implausible. Chairs are no doubt built by humans, but they are built *out of* wood and similar materials that are not themselves built; more generally *all* our building practices take place within a surrounding natural context—a context that makes building possible but is not itself built. Reification occurs when entities that are in fact socially constructed appear as independent and even "natural" Things; but some objects *are* independent and natural, and (one might be inclined to argue) it makes no sense to suggest that *those* entities too are our products.

Yet to say this seems to require distinguishing between "nature" and the human world in a way that significantly limits the range of applicability of the philosophy of practice, for the dualism it introduces looks suspiciously like the old one between subject and object that it was supposed to overcome. The claim that the mysteries that mislead theories to mysticism can only be resolved by appeal to human practice turns out to apply only to those mysteries having to do with the human world, and not to "nature." But the antinomies that the philosophy of practice was supposed to resolve had to do with the possibility of knowledge of the objective world itself, not simply some part of it. The empiricist views whose difficulties led to Kant's insight that knowledge must be active, then to Hegel's radicalization of that insight, and finally to Marx's materialist reformulation of it as a philosophy of practice began, after all, as an attempt to understand and explain the possibility of natural science; but if the philosophy of practice is explicitly denied applicability to nature then in fact those difficulties would seem still to remain. Theory and practice, interpretation and change, objective world and subjective/human one, mind and body, is and ought—all the dualisms that the philosophy of practice was supposed to dissolve—now come rushing back. And materialism itself is put at risk: the idea of grafting a strong distinction between nature and the human onto the monistic metaphysics that materialism asserts seems awkward at best: Aren't humans part of nature?

Feenberg is deeply aware of and concerned about this problem (which has important implications for any discussion of the social character and meaning of contemporary science and technology) and comes back to it repeatedly throughout his book. He examines both Marx's and Lukács's views about it in some detail and offers sophisticated and helpful readings of their discussions. Although sympathetic to the more radical version of the philosophy of practice from which "nature" is not excluded, ultimately he pulls back from this conclusion, which (in both the original book and the new version) he calls "rigorously consistent and obviously absurd."[20] The key tenets of the philosophy of practice, he repeatedly suggests, do not apply to nature. As opposed to systems such as the economy, technology, and bureaucratic administration, he writes in his discussion of Lukács's view of reification—systems whose reification can in principle be overcome by self-conscious practice—the realm of nature "is *essentially* reified and knowledge of it is destined to remain permanently 'contemplative.'"[21]

I don't agree with Feenberg, and want briefly to explain why. I don't want to claim that nature is a social product, which surely sounds odd, nor (as the philosophy of practice would seem to require) that self-conscious practices regarding nature would lead to nature itself changing, which seems odder: instead I want to question the significance and validity of the concept of "nature" itself.[22] To talk of nature is to talk of a world independent of human beings, an objective world separate from them that confronts them as alien, a world in which they act and whose laws they must obey and which they surely cannot change. Such a world is *of course* "essentially reified," by definition. The question is whether "nature" so defined correctly names the world we actually inhabit, or whether instead that name is being misapplied to the very world that the philosophy of practice wants to unmask as the reified product of practice. The real question is: How many "worlds" are there? Despite various attempts to deny it, Feenberg's view ultimately depends on a dualism that sees reality as divided into "spheres" or "realms" or "domains," with history or the social or the human being on one side and nature on the other. "There is a realm in which consciousness is practice," he writes, "in which we can transform our objects by becoming socially self-conscious," but alongside it "there is another realm in which our action will always be contemplative, that is, technical. The first realm is society, the second is nature."[23] Elsewhere he talks of history as a "special sphere" in which humans "are actually able to transform the objects on which [they] act," and describes Marx as intending to "subordinate" the "apparently humanly indifferent sphere" of nature to that special one.[24] "Unlike nature," he writes at another point, "history is the product of human action," and describes history as "the only domain in which to find a practice that can affect... the very essence of the phenomena."[25] But here his failure to understand practice specifically as *labor*, which is to say as *a process in which the human and the non-human are intertwined*, comes home to roost. For the notion of labor cuts across all these dualisms: Labor produces a world of *objective things* that are *fundamentally human*, while also producing a *social order* that is *entirely objective and material.*

There is only one world, not two. The passive materialism that Marx criticizes in the *Theses on Feuerbach* understands this one world as simply "nature," while the classical German idealists identify it with subjectivity. But the activist materialism of the philosophy of practice insists that labor produces a world that is *both "natural" and human*—a world that is fully material but is produced ("constituted") by human action, by labor. This means that everything Feenberg sees as "social"—history, culture, the economy, etc.—is *completely material*, and that it is a mistake to understand it as a human element somehow *added on* to a natural one. The economy, for instance, is not an abstraction from the various concrete makings and exchangings that take place in the marketplace: It simply *is* those makings and exchangings. Nor is the political system anything beyond the actual practices of voting, meeting, deciding, etc., of the people taking part in it. But by the same token "nature" is not something that exists *before* or *beneath* the practices humans engage in, but rather is simply *the world we inhabit*, the world in which those practices take place and which consists of things that those practices have helped to build.[26] But then (as I've argued elsewhere) rather than calling it "nature," a word infamous anyway for its complexity and ambiguity and especially for the antinomies it so easily engenders, we might be better off dropping that word and speaking of the "environment"—meaning the world that actually environs us, the one we actually inhabit, which nowadays anyway is a *built* world, the product of our labor.[27]

The "environment" consists of *all* the things that our labor has built: the artifacts and buildings that surround us, the social and economic structures and ideals that organize our lives, the technology we employ and its effects on us, our gender and class relations, our educational processes and medical procedures, our political institutions and our artistic movements, our history and our geology, our biology and our physics. *It's all one world*, and it is the product of our practices. Once this is recognized, the idea that there are different "realms" or "spheres" that differ either ontologically or even in terms of how we come to know them becomes untenable. All of reality is generated through practice, which means that practice comes *before* any distinction between human and natural, between subject and object, between idea and matter. In fact—and this is crucial—these distinctions are themselves *symptoms* of reification, deriving from a set of practices that do not know themselves as such and therefore fail to see the human character of the world those practices produce.

Feenberg goes part way toward acknowledging the point I'm making, coming close to admitting that our relationship to "nature" is always mediated through our practices, and that even the "contemplative" approach characteristic of the natural sciences depends upon a set of practices as well.[28] (These would include, for instance, practices of experimentation, of measuring and calculating, of instrument-building and so forth.) But he does not fully accept the consequence that seems to me to follow from this concession, which is that if the word "nature" *means* a world independent of and prior to our practices, *we have no access to it*, even scientifically, and so it is no longer clear in what sense we can call nature a "realm" of the world we actually inhabit. Once the practical character even of natural scientific knowledge is recognized, what that knowledge calls "nature" seems to turn out to be as much part of the "built" world we inhabit as are any of the other products of human labor—not *essentially* reified, but rather reified in the same sense as all those other products, which is to say open to an unmasking that reveals the human role in their construction.

Feenberg insists instead that it is on this last point that "nature" differs from the social. Recognition of the constructed character of social phenomena leads directly toward a *change* in the phenomena, he argues, while noticing the practical processes that underlie our understanding of nature has no such result: "dereifying consciousness of the scientific construction of nature does not necessarily alter the 'facts' of nature itself."[29] Elsewhere he writes that "knowledge of nature simply does not respond to self-consciousness."[30] The idea here seems to be that the discovery of the social practices underlying our grasp of certain elements of the world (the "natural" ones) need not automatically lead to a change in those practices and therefore might leave our understanding of those elements unaltered as well. Now if this were true, one might indeed want to define "nature" as that part of the world possessing this property; but Feenberg offers no argument whatsoever to show it to be coextensive with the nature investigated by the natural sciences, or more generally with what we usually think of when we talk of nature. (Nor, for that matter, does he show that there might not be aspects of the social that possess this property as well.) Instead he seems simply to assert that nature (as ordinarily understood) does possess this property (and that history does not), without explaining how that assertion could be justified.

But second, the very idea that we could recognize that something we thought was independent is actually produced through our practices without that producing a change in those practices reproduces the very dualism between thought and world (theory and practice) that the philosophy of practice puts into question. To "know" or "recognize" something that was previously hidden is itself a practice, not merely an event inside one's head; to realize that what once seemed to be an independent Thing is in fact the product of one's own practices *is* to engage in different practices, and so will produce a different object as well. This is not so difficult to believe if we think of the material world around us not as "nature" but rather as built. To recognize that we are communally responsible for the world that surrounds us, that we have built it, would almost certainly lead to the building of different objects: it is hard to imagine any self-conscious community choosing the environment of ugliness, dehumanization, and impending climate change that surrounds us today. But by "nature" Feenberg has something else in mind, although he never makes it clear exactly what. He seems to mean something like the object of inquiry of the natural sciences; and what he finds implausible is the idea that recognizing the practical character of scientific investigation could possibly lead to a change in the scientific "facts." Yet again he gives no evidence for this claim. And a geology that has to consider and investigate the Anthropocene, a psychology that has to face its own gender biases, a biology that works by way of genetic engineering and that increasingly produces chimeras as objects of study, and for that matter a physics among whose tenets is the notion that the character of basic elements of the world changes depending on whether they are being observed—all of these seem to suggest that recognizing the constructed character of the objects of natural scientific inquiry might indeed play a role in reconceptualizing those objects and interacting with them in different ways.

I find it genuinely difficult to understand exactly what Feenberg thinks about the ontological status of "nature" and its relation to practice, or what it means to call it "essentially reified." He argues at one point that "since nature as a system or totality does not depend on the unconsciousness of the practices in which it is understood, self-consciousness does not overthrow its reified form of objectivity although some results of scientific research may indeed be overthrown."[31] I'm not sure what nature's "reified form of objectivity" means here. A "form of objectivity" (Feenberg's translation of Lukács's *Gegenstandlichkeitsform*) seems to be the form in which objectivity appears in a given social order, which is to say given the practices that social order engages in.[32] To say that it is a *reified* such form would then seem to mean that those practices are not recognized as such. But then once those practices become self-conscious it is hard to see how the form would remain reified, even under Feenberg's questionable assumption that the practices themselves would remain unchanged. How could a self-conscious set of practices *still* confront a reified (*essentially* reified?) form of objectivity?

Similarly, in a later section responding to objections, Feenberg gives a confusing account of reification according to which "a reified object is one that has the *form* of an *independent fact governed by laws*," and glosses "independent" as meaning that "whatever social processes involved in its institution are occluded."[33] But then he explains that "to argue that certain types of objects are 'really' or 'essentially' reified does not mean that no such processes underlie their existence, but that their reified form is unalterable in practice."[34] If the social processes are self-conscious, however, and know themselves as "instituting" the object, then won't the "occlusion," and with it the reification, disappear? Or is Feenberg now making the stronger claim that the social processes involved in the institution of nature cannot become known (and so the processes cannot become self-conscious) even though they exist? (And even though apparently theoreticians of reification such as himself somehow do know them?)

One gets the impression that Feenberg at bottom is committed to the view that nature *really is* a realm independent of practice and that this is what he means when he calls it "essentially" reified. When challenged he acknowledges that our only access to that realm is a practical one, but still he wants to insist that there is a hardness and reality to that realm that is "prior" to practice, and that to deny this is to come too close to an idealism whereby humans somehow produce the world ex nihilo. But again it is the failure to grasp practice as real material labor that causes the problem, it seems to me. There is no question that in our practices we experience a resistance and hardness to the world that is not itself produced *by* practice. But it is a mistake to hypostasize—or to be more blunt, to *reify*—that hardness into something called "nature" that *causes* or *evinces* it. Doing so reinstates a dualism where humans are "outside" of nature and then attempt to achieve their goals by engaging in practices "upon" it. *Practices are material*: They take place in the (one) real world. And so *of course* they involve the experience of resistance and hardness. *That's what it means to be a practice*—to involve effort, to face resistance, to require the expenditure of energy, and indeed sometimes to fail. It is not that once a practice gets into the real world of "nature" it finds itself thwarted by the latter's reality: instead the point is that to be a practice is *already to be real* and so to involve difficulty and effort *by definition*. The difference, one might say, between practice and "theory" is exactly that the latter does not involve such difficulty. If we think of practice as the "application of theory to the world," then the moment of resistance will seem to be a characteristic of the world, and so will seem to be something independent of and prior to practice. But if we understand practice as material activity in which both the ideal world of theory and the real world surrounding us come to be what they are, then the idea that either one is independent of practice no longer makes any sense—not because practice somehow magically constitutes reality ex nihilo, but because practice is itself real.

An important point Feenberg repeatedly emphasizes is that the overcoming of reification should not be understood as a single transformative moment after which everything will be different and reification will disappear.[35] Instead he emphasizes that it has to be seen as a *continuing process of mediation* and indeed a potentially unending one. "Reification is … not the 'opposite' of dialectics," he writes, "but a moment in it."[36] Dereifying practices always occur within an objective context and under objective constraints, and in that context such constraints necessarily appear as Things independent of practice, which is to say they are reified. In this sense total dereification is impossible, except as a regulative ideal.[37] And as Feenberg points out, this means too that dereifying practices never create a world ex nihilo, but rather always on the ground of previous reifications. "Socialism is a reorganization of the society around a dialectical mediation of the reified capitalist inheritance," he writes; "reification is never completely eliminated but [rather] is repeatedly overcome in an 'unbroken alternation of ossification, contradiction and movement.'"[38]

The key idea here is that "proletarian practice does not create social reality, but rather mediates it."[39] But the dialectical point that all practices are mediations of a previous reality which, *relative to the practice*, serves as a Thing "prior" to practice is not the same as the claim that there is some Thing or realm called "nature" prior to *all* practices. The point that Feenberg is making here is in a way more obvious than he recognizes, once practice is understood as labor (as material) and not in terms of the obscure notion of historical practices that produce "social reality." Laboring practices always take place in material contexts and are subject to material constraints. The idea of labor producing something ex nihilo is ridiculous: labor always has a matter on which it labors. But that matter may have been, and almost always is, the product of *previous* labor. No notion of a "nature" that necessarily underlies all practice is required here. The constraints under which "social reality" is changed are of the same order as those under which all labor operates: they are simply the constraints of *reality*.

V

The question about the status of nature in the philosophy of practice is particularly significant in the contemporary context, where environmental problems from pollution to climate change to the protection of wilderness are very much on the agenda in a way that they were not for Marx and Lukács. In the latter chapters of his book, Feenberg provides an insightful discussion of the Frankfurt School, focusing on Adorno and especially on Feenberg's mentor Marcuse, and emphasizing how those thinkers reformulated the question of nature (and of science) in ways more relevant to environmental issues. Still there too, I would argue, his mistaken views about nature lead to problematic conclusions, including too quick of an appropriation of certain characteristic Frankfurt School approaches.

The key idea introduced by first-generation Frankfurt thinkers, as Feenberg notes, is the "domination of nature," which he says they see as "the central issue of the twentieth century."[40] Whereas earlier Marxism focused only on the domination of human beings by other human beings under capitalism, he asserts neither Marx nor Lukács "realized that insofar as the dominated human being is reduced to a natural object, all of nature is implicated in the social critique."[41] Capitalism treats humans and nature alike as objects to be manipulated, organized as it is by a principle of rationalization operative both in the economic sphere and in that of natural science. "Disenchanted" in the Weberian sense, nature appears under capitalism as mere matter available for human control, empty of meaning or ethical significance. This is supposed to take place in the name of human progress, but in fact—since human beings themselves are natural—ends up turning humans too into raw material for capitalist rationalization. Such is the "dialectic of enlightenment"

Adorno and Horkheimer trace in their book by the same name. The progress of science and rationality leads to a "fully enlightened earth [that] radiates disaster triumphant," a disaster that they identified with the political catastrophes of the mid-twentieth century but that surely could be seen to include the ecological ones later decades would come to know as well.[42]

The Frankfurt School view involves a critique of natural science that goes far beyond the one implicit in Lukács and the early Marx, Feenberg shows. He interestingly interprets the "critical method of science" and its universality not merely in terms of its commitment to empirical evidence and fallibilism but also in terms of the fact that it "systematically negates lived experience," removing the humanly meaningful elements of the natural world and thereby "authoriz[ing] the exploitation of nature as raw material."[43] Adorno and Horkheimer are famously pessimistic about the possibility of avoiding the fatal dialectic they trace; faced with a choice among fascism, Stalinism, and the culture industry of the West, they see no serious non-dominative alternative in the offing, no way to imagine a "liberation of nature." But Marcuse, Feenberg suggests, writing somewhat later, finds in the "new sensibility" of the counterculture and the New Left hints of a different approach involving what he calls "the recovery of the life-enhancing forces in nature." For nature, Marcuse writes, "has a dimension *beyond* labor, a symbol of beauty, of tranquility, of a non-repressive order."[44] The idea would be of an approach to nature that treats it not as "raw material" but as itself in need of liberation—that treats it, in a word, as another subject.

The line of argument here is familiar, yet from the point of view of the philosophy of practice it faces real difficulties, beginning with its unexamined use of the concept of "nature." The idea that in certain sorts of scientifically directed practices humans "dominate" nature while other "non-repressive" practices allow the "recovery" of its "life-enhancing forces" and hence make possible its "liberation" again implicitly appeals to the kind of dualism that the philosophy of practice wants to overcome. It divides the world in two, with humans on the one side and nature, described as implicitly another subject, on the other. When humans violate the boundary by attempting to manipulate and control nature, they fail to show it the proper respect and instead treat it as "mere matter" for their use. The transformation of nature looks like an illegitimate assault by one subject on the body of another.

But if we recognize that there is only one world, not two, and that it is neither a human world of limitless will to power nor a world of "nature" from which the human being is excluded, then we see that human practice is not, and never could be, the imposition of the human onto the natural, but rather that engaging in practice is simply *what it is to be a human being*. And to see the world as something we transform is not to see us as "dominating" it but rather recognizing ourselves as living active creatures in an environment that shapes us as we shape it. "Nature" does not appear as "mere matter" here—in fact, of course, "nature" doesn't appear at all. To understand the environment we inhabit as one we have helped to produce is not to fail to respect it but rather simply to understand that our entanglement in it is complete. To see the world as the product of our doings is to recognize our *responsibility* for it—both in the sense that it would not be what it is without us and also in the sense that we are *normatively responsible* for what it is, in the sense that if we find ourselves living (as we do, sadly) in a bad and ugly and dangerous world *this is our fault*, and that we are under a moral obligation to make the world a better one—which means to make our practices better ones, making better choices (and more communally self-conscious choices) about what practices we want to engage in based on our best expectations about what those practices are likely to bring about.

There is no "dimension" of nature "beyond labor," first because *there is no nature* if "nature" means a world that we have not already changed, and second because to talk this way is to treat labor the way capitalism does, as a kind of painful toil imposed upon us by an external force that produces an ugly and harmful world in which we cannot recognize ourselves. But labor just means practice, and so it includes all the practices Marcuse and Feenberg want to support: creative ones, playful ones, practices oriented toward beauty, toward the ornamentation of life and the flourishing of humans and other living organisms. An environmentalism that thinks its goal is to avoid dominating nature is one that sees human activity in the world as intrinsically repressive and harmful, and thus is one that requires us to be passive, to view the world as something independent of us whose structure we must acknowledge and whose requirements and laws we must obey; it is marked by a kind of fear that if we fail to obey them—if we try to "force" nature to do what we want instead of allowing it its own autonomy—it will take its revenge on us, via global warming or other mass catastrophes. But in the context of the philosophy of practice the call for such a passivity (which is in any case impossible) sounds like nothing other than what Lukács called the "contemplative attitude" that views a reified reality as something outside of human control—like the commodities that seem to have all the power in Marx's account of capitalism, despite being themselves the product of human labor. To recognize that the world that surrounds us is *not* something other than us—neither the world of nature nor the world of history—and to see our role in it as an *active* one, makes possible an environmentalism whose goal is a human community that acknowledges its responsibility for the world and takes that responsibility seriously: one whose citizens decide together what practices they will engage in, and what environment they want to inhabit, and who do not allow those decisions to be made for them by putatively external processes, whether those be the processes of "nature" or of the market.

Notes

1. Andrew Feenberg, *Lukács, Marx, and the Sources of Critical Theory* (Totowa, NJ: Rowman and Littlefield, 1981).
2. Andrew Feenberg, *The Philosophy of Praxis: Marx, Lukács, and the Frankfurt School* (London: Verso, 2014).
3. Ibid., 3.
4. Karl Marx and Frederick Engels, *Collected Works,* vol. 3 (New York: International Publishers, 1976), 4.
5. Feenberg, *The Philosophy of Praxis,* 100.
6. *Ibid.*, 102–105.
7. Marx/Engels, *Collected Works,* vol. 3, 3. Emphasis in original. See Karl Marx and Frederick Engels, *Ausgewählte Werke* (Moscow, Verlag Progress, 1972), 26.
8. Marx/Engels, *Collected Works,* vol. 3, 3. Emphasis in original.
9. "You didn't build that," Barack Obama once said (and was castigated for saying).
10. Feenberg, *The Philosophy of Praxis,* 44.
11. Ibid., 45.
12. Ibid., 44.
13. Ibid.
14. I'll argue below, though, that they are not examples of human relations to "nature." I'd rather say they are relations to the world.
15. Feenberg, *The Philosophy of Praxis,* 45.
16. Karl Marx, *Capital,* vol. 1, Ben Fowkes, transl. (New York: Vintage Books, 1977), 165.

17. See Steven Vogel, *Thinking Like a Mall: Environmental Philosophy after the End of Nature* (Cambridge, MA: MIT Press, 2015), Chap. 7.
18. See, e.g., Feenberg, *The Philosophy of Praxis*, 110–117.
19. Ibid., 155, 170–174.
20. Ibid., 129.
21. Ibid., 136–137.
22. See Vogel, *Thinking Like a Mall*, Chap. 1.
23. Feenberg, *The Philosophy of Praxis*, 135.
24. Ibid., 4.
25. Ibid., 111.
26. *Helped* to build: of course other organisms, and other things, played important roles as well. But this fact does not turn the environment into anything "natural" in the sense of somehow being separate from humans.
27. See Vogel, *Thinking Like a Mall*, 42–43.
28. See, e.g., Feenberg, *The Philosophy of Praxis*, 135–136.
29. Ibid., 136.
30. Ibid., 147.
31. Ibid., 141.
32. Ibid., 73–77.
33. Ibid., 147. Emphasis in original.
34. Ibid.
35. See Ibid., 112–119.
36. Ibid., 114. (The remark is repeated for some reason on p. 143.)
37. See Ibid., 117–119; 133–134.
38. Ibid., 134. (The quoted phrase is from *History and Class Consciousness*.)
39. Ibid., 133. See also Ibid., 123.
40. Ibid., 152.
41. Ibid., 156.
42. Theodor Adorno and Max Horkheimer, *Dialectic of Enlightenment*, John Cumming, transl. (New York, Seabury Press, 1972), 3.
43. Feenberg, *The Philosophy of Praxis*, 144–145.
44. See Ibid., 183. The quotations are from Marcuse's essays "Ecology and Revolution" and "Nature and Revolution."

Bibliography

Adorno, Theodor and Max Horkheimer. *Dialectic of Enlightenment*. Translated by John Cumming. New York: Seabury Press, 1972.

Feenberg, Andrew. *Lukács, Marx, and the Sources of Critical Theory*. Totowa, NJ: Rowman and Littlefield, 1981.

———. *The Philosophy of Praxis: Marx, Lukács, and the Frankfurt School*. London: Verso, 2014.

Marx, Karl. *Capital,* vol. 1. Translated by Ben Fowkes. New York: Vintage Books, 1977.

Marx, Karl, and Frederick Engels. *Ausgewählte Werke.* Moscow: Verlag Progress, 1972.

———. *Collected Works,* vol. 3. New York: International Publishers, 1976.

Vogel, Steven. *Thinking Like a Mall: Environmental Philosophy after the End of Nature.* Cambridge, MA: MIT Press, 2015.

CHAPTER 3

Philosophy of Praxis or Philosophical Anthropology? Andrew Feenberg and Axel Honneth on Lukács's Theory of Reification

Konstantinos Kavoulakos

The concept of reification, as it is formulated in Lukács's legendary book *History and class consciousness* (1923), constitutes one of the cornerstones of the tradition of critical theory.[1] The interest in Lukács's theory revived during the students' uprising of the 1960s. But after the decline of the New Left it was limited to the narrow circle of those theorists who remained faithful to the old critical theory's line of thought. Today, after a period of partial eclipse, the discussion of the classic formulation of this concept in Lukács's early Marxist work seems to have acquired new impetus, due to the theoretical interventions of two theorists of the critical theory tradition, namely Andrew Feenberg and Axel Honneth.

Both of them share the view that a concept of reification can form the basis of a theory of contemporary social pathologies, as it allows the theoretical disclosure of social phenomena, which cannot be adequately theorized by established political philosophies or by theories of justice. However, in their efforts to update the Lukácsian concept they choose

K. Kavoulakos (✉)
Department of Philosophy and Social Studies, University of Crete, Rethymno, Greece

D.P. Arnold and A. Michel (eds.), *Critical Theory and the Thought of Andrew Feenberg*, DOI 10.1007/978-3-319-57897-2_3

different, even contrasting ways, of interpreting it. In this chapter I will examine their respective attempts in order to determine which view could be considered more appropriate for its actualization at our current social and political juncture.

I will start with the interpretation of Lukács's theory of reification as proposed by Feenberg within the frame of the "philosophy of praxis" (Sect. "Reification as the Central Concept of the Philosophy of Praxis"). Subsequently, I will reconstruct Honneth's updated version of the concept in "recognition-theoretical" terms (Sect. "Reification as Forgetfulness of Recognition") to prepare the ground for criticizing this view from the perspective of Feenberg and Lukács's himself (Sect. "A Reified Concept of Reification"). I will conclude with some remarks on the actuality of Feenberg's interpretation of the Lukácsian concept of reification (Sect. "Reification and Dereification Today").

Reification as the Central Concept of the Philosophy of Praxis

Andrew Feenberg has not ceased to reflect on the question concerning the adequate philosophical foundations of contemporary critical theory since the appearance of his first book *Lukács, Marx and the Sources of Critical Theory* in 1981. Its new, radically revised edition entitled *The Philosophy of Praxis. Marx, Lukács and the Frankfurt School* (2014) is a clear indication of his intellectual persistence. In the *Philosophy of Praxis*, the reflections on science and technology Feenberg has developed during the last thirty years provide a firm theoretical basis for his historical reconstructions. His aim is to reappraise the theoretical tradition of "philosophy of praxis" from the standpoint of a critically reformulated Marcusian critique of scientific rationality and technology, informed by a reappropriation of Lukács's theory of reification. For Feenberg the importance of such a survey becomes crucial in view of the current crisis of modernity's rational foundations.[2]

According to Feenberg the philosophy of praxis takes its cues from a metacritique of idealist philosophy to show that the latter's main concepts and their antinomial relations are nothing more than philosophical sublimations of real moments of social life[3]—this was Marx's method in demystifying classical German philosophy. Subsequently the philosophy of praxis aims to transcend the limits of idealist philosophy by shedding light on the only real possibility for resolving philosophical antinomies, that is, overcoming them practically through social change.[4]

As an intellectual heir of idealist philosophy, the philosophy of praxis retains some of the basic ideas of the Hegelian approach in a demystified form. Firstly, it aims at transcending the epistemological standpoint by use of an *ontological* notion of the primordial unity between man and the world.[5] Secondly, human action takes on various (inadequate) historical forms; thus, *history* represents the ultimate reality.[6] Thirdly, an essential element of this approach is the *dialectical relation* between subject and object, hence also between form and content, mind and body, value and fact, etc.[7] However, for the philosophy of praxis the dialectical unity of the opposites cannot be restored in mere thought—it can be accomplished only through the application of the "principle of practice" in revolutionary action. In this sense, revolution represents the "realization of philosophy."[8]

In his early philosophical work Marx conceptualized labor and the human senses as the main ontological categories of the mediation between man and the world, while criticizing formalism for concealing the power of life's content to induce changes in the social forms of its mediation.[9] Lukács's early Marxist philosophy is presented by Feenberg as a more elaborate version of the philosophy of praxis.[10] According to Feenberg's reading, Lukács's great contribution is to the *cultural* understanding of capitalism and the possibilities for overthrowing it.[11]

Lukács's theory of reification, which established him as the founder of twentieth-century Hegelian Marxism, rests on the idea that cultural change can be explicated in terms of the practical alteration of a central, historically contingent "form of objectivity."[12] Feenberg recognizes the Neo-Kantian origins of this concept as well as the fact that, contrary to the Neo-Kantians, for Lukács the form of objectivity becomes a social-historical—and therefore historically changeable—category of both the epistemic and the practical spheres of experience and human life. The capitalist form of objectivity can be understood as a cultural pattern that forms the relation between subject and object, based on calculative-instrumental rationality.[13] Through these modifications the form of objectivity becomes the conceptual basis for Lukács's metacritique of the established forms of cognition and action.

On the basis of the concept of the form of objectivity, theory turns its attention toward the social totality and its unitary organization of epistemic and practical activities. It allows Lukács to formulate a powerful theoretical alternative to mechanistic Marxism through a dialectical-holistic theory of modern culture, elaborated in terms of the a priori of social-historical forms and cultural patterns.[14] At the same time, this theory acknowledges the root of the crisis and fragmentation of the social totality, since the dominant form of objectivity proves to be inadequate for and in conflict with the content it forms and organizes. Indeed, the content "overflows" the form and throws its adequacy into question. Therefore, the world shaped by this form is permanently shaken by successively erupting crises that result from the unresolved conflict between form and content. Because of its inadequacy, form has to be imposed on material content, the otherness of which is violated and suppressed.[15]

By emphasizing the concept of the form of objectivity and its cultural understanding, Feenberg is able to eschew a frequently made mistake in the interpretation of the theory of reification: the error of identifying objectification and reification. In my view, explicitly making this distinction would aid us in avoiding such a misunderstanding—but even implicitly presupposing it, as Feenberg does, can prove to be helpful. For him, reification is the consequence of the generalized social application of formal rationality as a cultural pattern.[16] It represents the phenomenon in which the social forms that are engaged in everyday practice appear as "natural facts," as an unalterable condition of human experience and praxis in general. According to Feenberg's cogent interpretation, this is what Lukács meant when he explicated reification as "treating social relations as things."[17]

Furthermore, although the inadequate form of social mediation is naturalized and its domination is fostered and perpetuated, it is precisely its inadequacy that provides a possible solution, since it motivates the proletariat as a *collective subject* to change the social form of objectivity through revolutionary praxis. Because the content of life is violated by its social form, the possibility of a conscious "fighting back" arises:

> Life overflows rationalization in every direction and comes back to haunt the rationalized domains in the course of class struggles. These struggles reveal the human basis of the society that has been shaped but also repressed by the reified forms. This provokes reflexive processes unknown in nature. Human lives considered as content of the reified forms have an independent power not just to violate expectations but also to understand themselves as doing so. This resistant self-understanding constitutes the core of what Lukács, following Marx, calls class consciousness.[18]

In a heretical interpretation that opposes the superficial readings of Lukács's early Marxism proposed by Adorno, Habermas, and even old Lukács himself,[19] Feenberg demonstrates why revolutionary change, as described in *History and Class Consciousness*, cannot be comprehended as a singular act of the absolute production of a new, harmonious, and transparent reality. Revolutionary change must be understood as a long process of struggles, a process of consecutive precarious dialectical mediations with no final solution. In this sense dereification involves the continuous removal of institutional and cultural impediments to an open process of emancipative social change rather than an idealist/theological creation of all objectivity by a mythical collective subject.[20]

Feenberg also works out how Lukács saves his argument from reverting to metaphysics by exempting nature from dereification, thus retaining its otherness. However, this salvation causes a tension in his theory, since he admits that dereification can never concern the *totality* of the world. Lukács's "absolute historicism"[21] seems to meet a discernible limit here, which implies the apparent danger of reverting to the antinomies of bourgeois thought.[22] But even if this internal conceptual problem did not occur, Lukács's philosophy of praxis would sooner or later face an "external problem" stemming from the changes in the social-historical situation.

After the frustration of the hopes for an immediate and successful continuation the socialist revolutions in Europe in the 1920s and 1930s, the Lukácsian mediation between theory and praxis through the constitution of the proletariat as a revolutionary subject no longer seemed to be possible. In fact, the experience of the failure of revolution constitutes the starting point of the Frankfurt School. Adorno and Horkheimer drew upon the Lukácsian critique of reification to formulate a totalizing critique of instrumental reason and domination over nature without the guarantees of a future reconciliation through social praxis.[23] In particular Adorno, who could not avoid implicitly presupposing the perspective of reconciliation with "fallen nature," systematically eschewed presenting it as a practical possibility.[24]

According to Feenberg, here lies the great significance of a third representative of the Frankfurt School, who fully recognized the emancipative potential of the revolt of the 1960s, namely Herbert Marcuse. Thus, with Marcuse "the Frankfurt School returns to its sources,"[25] namely to the initial versions of the philosophy of praxis as formulated in the works of Marx and Lukács. Marcuse's "original version of the philosophy of praxis"[26] is particularly significant for Feenberg's reappropriation of the Lukácsian theory of reification, which can also be described as a reading of Lukács's early Marxist work through the lenses of Marcuse.

In his early work Marcuse was already interested in an ontological interpretation of the relation between man and the world and in the diagnosis of its distortion by an inadequate, historically situated form of rationality in capitalist society. Following Adorno, he showed that "one-dimensional" rationality serves control over nature. Thus, in modern society nature is phenomenologically "projected" as a value-free matter to be rationally dominated. However, this rationality negates the various potentialities inherent in beings.[27] But at the same time Marcuse recognized in the protest of the New Left the traces of a new experience: an aesthetic relation to nature, capable of fostering a consciousness of the repressed potentialities, thereby anticipating a subversive culture of pacification, respect for otherness, and reconciliation with nature. In this way, Marcuse restored an essential element of the philosophy of praxis: The reference to the "real" subject of dereification and social change.[28]

However, Feenberg criticizes Marcuse's fascinating but problematic return to Marx's idea of a totally new science based on new experiential foundations. But at the same time he advocates Marcuse's idea of a new technology as much more convincing and reasonable. Technology rests not only on scientific rationality but also on *design* that mediates between scientific knowledge and its technical applications by forming a "technical code"—the latter can therefore incorporate life-affirming values to replace the current "formal bias" of technology that favors domination and control.[29] Marcuse's vision of a new technology can be further explicated through the critical theory of technology Feenberg has formulated during the last thirty years—a theory that aspires to contribute to a wide process of democratization of "technical politics."[30]

Marcuse's philosophy of praxis paved the way for a theoretical conceptualization of the continuing resistance against reification in an epoch of fragmented oppositional struggles: "The failure of proletarian revolution has not ended the struggle against reification but it has fragmented that struggle."[31] Thus, the new social movements make new demands opposing racial and gender discrimination, the socially disastrous effects of austerity policies, the ecologically dangerous, unrestricted exploitation of nature, and centralized, antidemocratic control over technological progress.[32] Such social movements "do not fulfill the conditions of revolution as Lukács explains them."[33] However, they correspond to the kind of practice Feenberg calls "cultural" or "transformative," because it aims at "establishing the framework of meaning within which activity goes on."[34]

As has been noted, "the 'philosophy of praxis'...is Feenberg's attempt to recognize technology as self-alienated social practice, or to use Lukács's term, 'reified' action that engenders political irresponsibility, the false naturalization or hypostatization of activity that could be changed."[35] In fact, for Feenberg, the Lukácsian notion of *mediation* can help us understand the new movements around technical politics as a moment in the dialectics of reification and dereification.[36] Resistance to reification may still be politically weak and scattered; however, it operates on a deeper level, changing the capitalist a priori of everyday experience, drawing upon the cultural resources of a "new sensibility" informed by aesthetic experience, as Marcuse would like to have it. The significant contribution of Feenberg's work to the debate on contemporary critical theory is the critical reappropriation of an important theoretical tradition of modernity and the reconstruction of reification as its central concept.

Reification as Forgetfulness of Recognition

Contrary to Feenberg's appropriation of the Lukácsian concept of reification, the interpretation of it offered by Honneth in his Tanner Lectures (2005), published under the title *Reification. A New Look at an Old Idea*,[37] has the unambiguous consequence of deradicalizing the concept. Following Habermas, Honneth connects Lukács's theory of reification with the outdated idealist philosophical presuppositions of the philosophy of the subject and the "paradigm of production."[38] Thus, according to Honneth's reconstruction of the "official version" of Lukács's theory, only that kind of practice in which the object can be thought of as a product of the subject is undistorted. For Honneth, such a theory represents a retreat to obsolete views of German Idealism's philosophy of identity.[39]

Therefore, Honneth searches for an "unofficial version" of the concept of reification. Contrary to the "official" version, the "unofficial version" proposes an "intersubjective attitude on the part of the subject"[40] as a model in the light of which reification appears as "an atrophied or distorted form of a more primordial and genuine form of praxis, in which humans take on an empathetic and engaged relationship toward themselves and their surroundings."[41]

For Honneth, what we need to cogently update the concept of reification is to radically change the methodological framework within which it is applied. Honneth constructs such a frame through two basic theoretical decisions: Firstly, he turns his attention to a problem that is, as he himself admits,[42] alien to Lukács's philosophical self-understanding–the problem of the *normative foundations* of his theory. According to Honneth, Lukács's critique is based on an *implicit anthropological criterion* that refers to the notion of "a more genuine or better form of human praxis."[43] Secondly, Honneth abandons the dialectical method of the theory of reification and takes recourse in the anthropological "primary–secondary" scheme to describe the relation between the original anthropological conditions of human life and the secondary infringements upon these or distortions of them in social life.

Honneth is right to connect the phenomenon of reification with "a form of praxis that is *structurally* false" and not with "a mere epistemic category mistake nor a form of moral misconduct."[44] However, the abandonment of the dialectical-holistic perspective leads to a series of significant alterations of the Lukácsian theory of reification: Instead of describing the consequences of the social implementation of a social–ontological category—namely the calculative form of objectivity—Honneth's concept of reification has the same trivial meaning one could give to it in everyday talk: Reification "designates a cognitive occurrence in which something that does not possess thing-like characteristics in itself (for example, something human) comes to be regarded as a thing."[45] It is ironic that interpreting him in this doubtful manner allows Honneth to ascribe to Lukács an "ontologizing everyday understanding of the concept of reification."[46]

In any case, Honneth credits Lukács with this simplified version of reification—as emotionally unengaged objectification of things and, above all, of human beings or even of oneself. In spite of the fact that in the beginning of his essay Honneth cites Lukács's definition, according to which reification means "that a relation between persons has taken on the character of a thing,"[47] Honneth does not pose the question why, for Lukács, it is "*a relation between persons*" that is being reified and not "something human"[48] in general.

Based on this peculiar interpretation of Lukács's theory of reification, Honneth develops a series of analyses about intersubjective recognition and its degeneration. He draws on Heidegger's concept of "care" (Sorge), Dewey's notion of a "qualitative experience," Cavell's primacy of recognition, and the emotional side of human communication, as well as on ontogenetic findings of contemporary developmental psychology, to unfold a multilayered argument about the primacy of recognition as the foundation for the constitution of the self and its relations to others. Recognition is (from a genetic as well as a logical standpoint) a presupposition of any objectifying attitude toward others, but also of the objectification implied in man's cognitive relation to the world.[49]

Honneth's "recognition-precedes-cognition claim"[50] gives him the opportunity to reformulate a critique often found in the relevant literature on Lukács, namely the charge that Lukács erroneously identifies reification with objectification in general, thus succumbing to an idealist way of thinking. By taking objectification as a cognitive procedure, as "objectifying thought,"[51] Honneth interprets the Lukácsian concept of reification as "a kind of mental habit or habitually ossified perspective" that takes the place of the original empathetic engagement of the subject with other humans and objects.[52]

This interpretation relies on a modification of the meaning of the Lukácsian concept of "contemplation." Honneth understands it as the "emotionally neutral," "detached stance" (*Teilnahmlosigkeit*) of a "neutral observer, psychically and existentially untouched by his surroundings," a stance of "indulgent, passive observation"[53]:

> Lukács consequently, understands "reification" to be a habit of mere contemplation and observation, in which one's natural surroundings, social environment, and personal characteristics come to be apprehended in a detached and emotionless manner—in short, as things.[54]

However, Lukács did not use the term "contemplative" to describe any psychological fact. Rather the term describes the passive stance of subjects who take the social-historical world as unalterable. This contrasts with the active attitude of subjects who intervene in the world in order to change its historical contingent necessities.[55] Be that as it may, for Honneth, Lukács's "fundamental idea"[56] is the following: Commodity exchange implies calculating prospective profits and therefore forming a "purely rational and emotionless stance."[57] This attitude becomes individuals' "second nature" insofar as socialization processes establish it as a "habit" in all fields of everyday life.[58] Like the concept of reification, the concept of "second nature" is interpreted by Honneth in the same sense we find it in everyday talk: "Human beings' 'second nature'"[59] is nothing but the sum of the socially established attitudes and behavioral patterns.[60]

Hence, reification has to be viewed as a *secondary* neutralization of the primary recognition and emotional engagement with other people and with objects in the world, accompanied by its permanent replacement by the ontogenetically derivative objectifying attitude of the cognitive subject as its "second nature."[61] Consequently, as Honneth takes it for granted that Lukács identifies reification with cognitive objectification, he feels justified in characterizing the former's theory of reification as "not sufficiently complex, not sufficiently abstract."[62]

In his eyes, Lukács's critique is too totalizing, precisely because it supposedly relies on the assumption that a total substitution of the objectifying attitude for the original recognition is possible.[63] However, for Honneth, such a conjecture would be absurd. It would entail the assumption that "every social innovation that requires that we neutralize our original act of recognition and make this neutralization institutionally permanent is a case of reification."[64] Social rationalization seems then to propel a totalization of reification in all fields of life. However, this supposition is at odds with Lukács's alleged need to simultaneously retain the assertion that the "original stance of empathetic engagement can never be lost—since, after all, it lies at the base of all social relations."[65]

For Honneth, what we can learn in view of the impasse of Lukács's "conceptual strategy" is that the concept of reification "must be understood differently than Lukács understands it in his own work."[66] To avoid the totalization of reification, which is supposedly implied in Lukács's understanding of it, Honneth suggests a distinction between two cognitive attitudes: one in which a "consciousness"[67] of "antecedent," original recognition is retained and one in which it is lost. Honneth connects the second kind of a cognitive attitude with the concept of the "forgetfulness of recognition."[68]

Reification as forgetfulness of recognition means then "that in the course of our acts of cognition, we lose our attentiveness to the fact that this cognition owes its existence to an antecedent act of recognition."[69] Honneth connects the causes of such a "reduction of attentiveness," on the one hand, with the one-sided focus of a subject on an aim that discards other, possibly more fundamental goals and, on the other hand, with the influence of social "prejudices" and "thought schemata" on its behavior.[70]

A Reified Concept of Reification

In what follows I would like to critique the essence of Honneth's "re-actualization" of the concept of reification, leaning on the views of Feenberg and Lukács himself. It is not by accident that Honneth's choice to detach the concept of reification from his dialectical-holistic theoretical framework finally reproduces what Lukács called the "antinomies of bourgeois thought."[71] Hence, one could consider this interpretation as a reification of the concept of reification. Let us briefly see why.

Although Honneth critiques Lukács's tendency toward idealism, a similar kind of a tendency to ignore the materiality of social structures can be traced in Honneth's own thought. As one commentator notes: "With his theory of recognition, Honneth grasps crucial dimensions of the normative order of capitalist social relations, but he does so at the expense of neglecting the material constitution of those relations."[72] It is characteristic that, whereas in the beginning of his essay he connects reification with a "structurally false form of praxis," as he unfolds his argumentation he shifts his critique from praxis to "knowledge," "cognition," or "consciousness."

In fact, as we have already seen, in the greatest part of his essay reification is interpreted as a mental "habit of perceiving" oneself "and the surrounding world as mere things and objects,"[73] as a "reduced attentiveness" for original recognition, as its "forgetfulness" and its replacement by a purely cognitive attitude, etc. While the material dimensions of the phenomenon of reification are absorbed by its mental dimension, the fact that the concept of "second nature" is interpreted in a way that totally eliminates Lukács's historical-materialist reference to the social mechanisms and their material laws, namely as the sum of man's acquired mental habits, no longer surprises us.

This idealist reduction of reification goes hand in hand with understanding it as an attitude *of the individual.* However, as Feenberg aptly notes in his critique of Honneth's understanding of reification, "for better or worse, the individual is of only marginal interest to Lukács. ... Lukács's discussion of reification focuses on social processes, specifically on what today we would call the dialectic of structure and agency."[74] Instead, Honneth is concerned with the *individual stance*, involved in interpersonal relations and the relation to oneself, and not with the dialectical interdependence between social institutions or cultural structures and human behavior. On the other hand, Honneth's blindness toward dialectical method leads him to the view that for Lukács the "expansion of commodity exchange" is the "social cause" of reification[75]—thus implying that his theory is reductive. Furthermore, he claims that Lukács explained the generalization of the phenomenon in bourgeois society by functionalist arguments or with the aid of Weberian views about the expansion of instrumental-rational action.[76]

In this way Honneth totally overlooks the form of objectivity as a concept of central importance for the reconstruction of the social and cultural *totality* that determines the frame of every causal or functional link between particular phenomena.[77] When Lukács speaks about the "function" of a "moment" within a social totality, he does not have contemporary functionalist theories in mind, but a theory of dialectical relations between particular elements of society. Besides, from his dialectical point of view, functionality contains its abrogation as a dialectical moment revealed in the phenomenon of crisis. In a similar way, Lukács integrates a Weberian understanding of instrumentally rational practice into a dialectical theory of social totality that is, in the final analysis, irrational.[78] Nevertheless, the widespread critique of Lukács's alleged economism is uncritically espoused by Honneth.

Even more questionable is the fact that in his reconstruction a classic formalist tendency toward separating different fields of social life spontaneously occurs—like the separation of the economy from interpersonal relations, ideology, politics, etc. The purpose of theoretically separating these fields is to blunt the "totalizing" edge of Lukács's critique and to locate spheres of social life (such as the capitalist market or the bureaucratic organization), in which "observing, detached behavior has a perfectly legitimate place," while its retention is required "for reasons of efficiency" in "highly developed societies."[79] With such authentically functionalist arguments Honneth discards Lukács's systems-theoretically "naïve," anti-capitalist orientation, which contemporary academic critical theory can hardly handle.

However, the problem with this theoretical perspective is not only that it acquits the capitalist economy and the bureaucratic organization of the state in advance by declaring them innocent of the phenomenon of reification. But rather that Honneth's explicit rejection of the mechanistic explanation and his implicit opposition to the dialectical-holistic explication fail to open the way for another model for illuminating the social roots of reification. A social explanation of "reification" of any kind seems to lie totally beyond the capabilities of Honneth's interpretation.[80] In this way even the social basis of the phenomenon of the emotionally neutral stance toward others, which Honneth understands as "reification," is obscured. This tendency is further strengthened due to the elaboration of reification through concepts of a philosophical anthropology that replaces Lukács's social-historical perspective.

Honneth's preference for anthropology alone would suffice to demonstrate the great distance that separates him from Lukács, who emphatically rejected philosophical anthropology as a kind of reified thought. For Lukács, an anthropological theory can be nothing but ahistorical.[81] Indeed, the antecedent, original recognition that is necessary for the constitution of human subjects can only be thought of as a universal, ahistorical, constant element of human existence that we find in all societies and epochs. Accordingly, its "forgetfulness" must be understood as a kind of constant anthropological trait or at least as a permanent possibility of an individual slip into "reification"—even more so, since it is not explained through a social cause.[82] This inevitably amounts to an ontologization or—in Lukács's terms—a reification of the phenomenon of reification.

Finally, even if it was possible to overcome the opposition between history and ahistorical anthropology in theory, it remains unclear how this theory would have consequences *for praxis*. One does not have to be a follower of the philosophy of praxis to accept that no critical theory can abandon the demand that it be connected with an emancipative social praxis without ceasing to be critical. But what exactly is the practice Honneth refers to in his theory of the forgetfulness of recognition? It is not by accident that he remains silent on that issue. After all it is meaningless to try to eliminate an ontological or anthropological characteristic of human existence.

As a radical change in the world in respect to reificaiton seems to be impossible, only the path toward human inwardness promises some relief. The fact that Honneth exclusively locates the phenomenon of reification on the level of "thought" or "cognition" motivates the interpretation that dereification has to be something equivalent, some kind of an "internal change" of the individual. Even if we are today—together with Feenberg—compelled to reduce our expectations from the "principle of praxis," which Lukács connected with revolution, Honneth's version of it is clearly too thin. By explicating dereification in terms of a "remembrance" of recognition, of the cultivation of a cognitive attitude with attentiveness to this fundamental anthropological presupposition, theory limits itself to formulating an ethical demand that the individual undergo an inner reform and achieve the correct mental stance in his or her interpersonal relations.

All the above-mentioned elements of a critique of Honneth's theory of reification strengthen the suggestion of its reifying tendency. In the final analysis they all have to do with Honneth's abandonment of Lukács's dialectical social-historical methodology. They thus lead to an abolition of its basic aim to demonstrate the *historically contingent character of capitalist social relations*. One can paraphrase Honneth's above-cited conclusion about Lukács's failure to teach us what reification is[83] in order to formulate an assessment of the achievements of his own theory of reification: What we can learn from Honneth's reconstruction is that a reinterpretation of Lukács's theory of reification must be constructed differently from the way Honneth understands it in his own work.[84] Indeed, Honneth's interpretation obscures Lukács's radical initial idea that human relations are reified as long as their historicality is forgotten and that it is precisely when this happens that they are transformed into an irresistible system of blind, compulsive, seemingly natural laws, in which people are reduced to passive observers of an independent, external process.

Reification and Dereification Today

Instead of trying to adjust the concept of reification to a theoretical framework so extrinsic to the one that gave birth to it, the enterprise of reactualizing it should better *take Lukács's initial idea seriously*. In this case we would need to counter the uncritical adoption of the all too often unjustified critiques of it we find in the relevant literature with a charitable and historically informed rereading. Only in this way is it possible to soberly judge whether a new clarification of the concept of reification has any diagnostic and political value today. This is the approach Andrew Feenberg adopted in the first edition of his book, in 1981.

However, this first edition was published in a transitional period for radical social critique: At that time the political and ideological impetus of the New Left was diminishing, leaving behind a number of scattered claims to liberation that still remain unsatisfied. Things were not better in the realm of theory. In the same year, 1981, Habermas's opus magnum *The Theory of Communicative Action* appeared in German—a book that established formalism's preponderance in the field of critical theory. Habermas's communicative reformulation of the old critical theory promised a realistic compromise between the instrumental imperatives of the system and the democratic-communicative demands of civil society that would lead to more individual and collective freedom. As a result, the claim that the old critical theory Feenberg was trying to update had been irreversibly overcome became a widespread conviction.

But since that time many things have changed. Habermas's theory was formulated in the same period that state-interventionism was harshly attacked by the rising ideological and political power of neoliberalism. The latter's dynamics proved to be strong enough to transform the world radically: The economic power of multinational corporations grew enormously due to the globalization process and the consequent deregulation of markets, while the communicative power of civil society decreased significantly. The new supranational economic elites systematically have taken advantage of the abolition of social control over national economies and of the unprecedented technological progress to establish their control over fragmented resistances occurring from below.

When Honneth first published his Tanner Lectures on reification in German in 2005, his argument seemed much more convincing when it dogmatically presupposed a politically neutral, unquestionable "efficacy" of instrumental action and its systems. This presupposition of the inescapability of "differentiated rational systems" in the modern world mirrors nothing but a widespread prejudice of the Habermasian critical theorists during the last thirty years.[85] It is a fashionable way of posing the question of the tension between capitalism and democracy in terms of their complementarity and not so much in terms of their opposition. But it is exactly this efficacy which has increasingly been called into question since the beginning of the global economic crisis in 2008—the very same year the English translation of Honneth's book appeared.

Today's multifaceted—economic, political, and ecological—crisis nourishes the feeling that we are not facing this or that particular problem but a *general crisis of the dominant system and its rationality as a whole*. Therefore, if in 2005 Honneth was in a position to claim that Lukács's radical critique of capitalist and bureaucratic rationality is outdated, today we are rather inclined to think that Honneth's moderate critique, which leaves the core principles of the global system untouched, is rapidly losing the ground to which it could be applied. The crisis and the new consciousness that grows out of it represent, as Lukács would like to have it, the first stage in a process of dereification through the practical resistance of those whose lives are being destroyed by the dominant social structures and principles.

Undoubtedly, this context renders an explication of crisis on the basis of a deeper inadequacy of these structures—like the one detected by the theory of reification in its classical formulation—more attractive and interesting. As there are still no cogent alternatives to capitalist globalization in sight, an urgent need for the theoretical elaboration of the negative experiences of the victims of austerity and ecological devastation arises. Feenberg's re-actualization of the philosophy of praxis meets this need on a philosophical level.

The process of transforming social institutions stands at the centre of Feenberg's considerations of the dialectical mediation between reified structures and the active resistance of men. Their dialectical interplay, which is missing from Honneth's theory, is an essential part of Lukács's holistic understanding of the interdependence between form and content. In his reconstruction Feenberg justifiably stresses the nonidentity of form and content and the relative independence of the latter, in the light of which reification must be viewed as a historically finite social phenomenon that can be abolished in its present form—firstly through the consciousness of its historical contingency, and secondly through conscious human intervention and modification of the fundamental form of objectivity and the relevant institutions that effect reification.

For Lukács dereification was never supposed to be a final praxis of the rational "production" of the object. As he noted in 1923, it has to be viewed as a dynamic process of "unbroken alternation of ossification, contradiction and movement."[86] Thus, within the Lukácsian perspective we should abandon the idea of a completely transparent and dereified social condition and envisage a future in which contradictory democratic control over all the fields of social life would at least be rendered possible.[87] This idea is central in Feenberg's rereading of Lukács's theory of reification. Of course, Feenberg connects the idea of a *democratic rationalization* first and foremost with politics on technology.[88] However, since his critique refers to the one-sided rationality of technical systems, it can obviously also be applied to the calculative rationality of the economic system. This would be an extension of his theory that is necessary in times of global economic crisis.

The concepts developed by Feenberg, such as the "formal bias" of rationality or "design," could be adjusted to a radical cultural critique of a growth economy, beyond the mainstream technocratic explanations of crisis that dominate the public sphere today.[89] Indeed, a "politics of dereification" could be based on a restored and updated concept of reification that refers to the contemporary tendency to form "rationalized systems," like markets, bureaucratic organizations, and technical systems of production. For Feenberg, such rationalized systems "are only partially differentiated from the life-world"; hence, "they are still subject to intervention and transformation from below."[90] Feenberg shows cogently that the philosophy of praxis does possess the theoretical potential to conceptually articulate the new cultural consciousness gradually formed by the new social movements and to delineate a positive emancipative perspective of present struggles. It thus can give rise to fruitful dialogue on the necessary re-radicalization of social critique in the face of the deep and multifaceted crisis of contemporary capitalism.

Notes

1. The most comprehensive reconstruction of Lukács's theory of reification can be found in Rüdiger Dannemann, *Das Prinzip Verdinglichung. Studie zur Philosophie Georg Lukács's* (Frankfurt/M.: Sendler Verlag, 1987).
2. Andrew Feenberg, *The Philosophy of Praxis. Marx, Lukács, and the Frankfurt School* (London, New York: Verso, 2014), vii. Agnes MacDonald aptly stresses the importance of this starting point for Feenberg's theoretical perspective in: "Andrew Feenberg: *The Philosophy of Praxis: Marx, Lukács, and the Frankfurt School*," *British Journal for the History of Philosophy* 22/4 (2014): 851–855.
3. Cf. Feenberg, *Philosophy of Praxis*, xiii–xiv, 9–10, 11–12.
4. Cf. ibid., 4, 16, 19.
5. Cf. ibid., 5.
6. Cf. ibid.
7. Cf. ibid., 14–15.
8. Cf. ibid., 19.
9. Cf. ibid., 32–33, 36–37.
10. Cf. ibid., xiv.
11. Cf. ibid., 41.
12. This central category of Lukács's theory of capitalism is firstly introduced in the introductory passage of the essay on "Reification and the Consciousness of the Proletariat," where Lukács explains that the commodity form constitutes the "model of all the forms of objectivity of bourgeois society together with all the forms of subjectivity corresponding to them." Georg Lukács, *History and Class Consciousness. Studies in Marxist Dialectics*, trans. Rodney Livingstone (London: Merlin Press, 1971), 83, translation altered.

13. Cf. Feenberg, *Philosophy of Praxis*, 65–66, 69, 73–77, 175–176.
14. Cf. ibid., 54–55, 63–66.
15. Cf. ibid., 67, 79–81.
16. Cf. ibid., 66–69.
17. Ibid., 62.
18. Ibid., 81.
19. See ibid., 125–129. Also see Feenberg's critique of Adorno's reading of Lukács in his "Reification and Its Critics," in *Georg Lukács Reconsidered. Critical Essays in Politics, Philosophy and Aesthetics*, ed. Michael J. Thompson (London/New York: Continuum), 172–194. I myself contested Habermas's, Wellmer's and Honneth's readings of Lukács in Konstantinos Kavoulakos, "Back to History? Reinterpreting Lukács's Early Marxist Work in Light of the Antinomies of Contemporary Critical Theory," in *Georg Lukács Reconsidered. Critical Essays in Politics, Philosophy and Aesthetics*, ed. Michael J. Thompson (London/New York: Continuum, 2011), 151–171.
20. Cf. Feenberg, *Philosophy of Praxis*, 113–119.
21. Ibid., 1.
22. Cf. ibid., 130–132, 134–136.
23. Cf. ibid., 151–158.
24. Cf. ibid., 167–171.
25. Ibid., 155.
26. Ibid., 155.
27. Cf. ibid., 175–182.
28. Cf. ibid., 172–174.
29. Cf. ibid., 185–186, 194–202.
30. Cf. Andrew Feenberg, "Lukács's Theory of Reification and Contemporary Social Movements," *Rethinking Marxism* 27/4 (2015): 490–507; Feenberg, *Philosophy of Praxis*, 212–214, 219–220.
31. Ibid., 203.
32. Cf. ibid., 219. It is questionable whether gender and racial discriminations—mentioned by Feenberg in this context—can be described as cases of reification, since their justification does not rely on the social generalization of formalist rationality but rather on prejudices–such as patriarchy, racism etc.—that have their roots in pre-capitalist epochs.
33. Ibid., 220.
34. Ibid., 203.
35. Chris Cutrone, "Andrew Feenberg, *The Philosophy of Praxis: Marx, Lukács and the Frankfurt School*," *Marx & Philosophy Review of Books*, February 14, 2015, http://marxandphilosophy.org.uk/reviewofbooks/reviews/2015/1524 (accessed 25 August 2015).
36. Cf. Feenberg, *Philosophy of Praxis*, 210–212.

37. Axel Honneth, *Reification. A New Look at an Old Idea*, ed. Martin Jay (Oxford/New York: Oxford University Press, 2008), translation of quotes at some points altered.
38. Unfortunately, this reading of Lukács, which has a long history within the Frankfurt School, continues to remain unchallenged by its commentators–see, e.g., Rahel Jaeggi, "Verdinglichung – ein aktueller Begriff," *Jahrbuch der Internationalen Georg-Lukács-Gesellschaft* 3 (1998/99): 70–71; Titus Stahl, "Verdinglichung und Herrschaft. Technikkritik als Kritik sozialer Praxis," in *Ding und Verdinglichung: Technik- und Sozialphilosophie nach Heidegger und der ·Kritischen Theorie*, ed. Hans Friesen, et al. (München: Wilhelm Fink, 2012), 303–304.
39. Cf. Honneth, *Reification*, 27.
40. Ibid.
41. Ibid.
42. Cf. ibid., 20.
43. Ibid., 26.
44. Ibid.
45. Ibid., 21. It has been correctly noted that Honneth takes reification in its "literal" sense. Cf. Andrew Feenberg, "Rethinking Reification," in *Georg Lukács: The Fundamental Dissonance of Dxistence*, ed. Timothy Bewes and Timothy Hall (London/New York: Continuum, 2011), 102; Timo Jütten, "What is Reification? A critique of Axel Honneth," *Inquiry: An Interdisciplinary Journal of Philosophy* 53/3 (2010): 236. Jütten pinpoints the internal contradiction that occurs between this literal meaning of the concept and Honneth's position that reification should not be interpreted as a deviation from a moral norm (ibid., 242–245).
46. Honneth, *Reification*, 21.
47. Ibid., 21; Lukács, *History*, 83.
48. Honneth, *Reification*, 21.
49. Cf. ibid., 28–52.
50. Jütten, "What is Reification?" 239–240.
51. Honneth, *Reification*, 54.
52. Cf. ibid., 53.
53. Ibid., 24.
54. Ibid., 25.
55. Feenberg sees that clearly in Feenberg, "Rethinking Reification," 104–105.
56. Honneth, *Reification*, 24.
57. Ibid., 25.
58. Cf. ibid., 25, 28.
59. Ibid., 23.
60. Cf. ibid., 23–25, 28, 32–33.

61. Cf. ibid., 54.
62. Ibid., 55.
63. Cf. ibid., 53.
64. Ibid., 54–55.
65. Ibid., 55.
66. Ibid.
67. Ibid., 56.
68. Cf. Ibid.
69. Ibid., 59.
70. Cf. ibid.
71. Lukács, *History*, 110.
72. Anita Chari, "Toward a Political Critique of Reification. Lukács, Honneth and the Aims of Critical Theory," *Philosophy and Social Criticism* 36/5 (2010): 599–600. Dirk Quadflieg detects a problem of neglecting the dialectics of subject and object in Habermas's and Honneth's accounts of reification and searches for a corrective in the early work of Hegel, seen through the lenses of the late Adorno. Cf. Dirk Quadflieg, "Zur Dialektik von Verdinglichung und Freiheit. Von Lukács zu Honneth – und zurück zu Hegel," *Deutsche Zeitschrift für Philosophie* 59/5 (2011): 708–714.
73. Honneth, *Reification*, 23.
74. Feenberg, "Rethinking Reification," 101. See also Timothy Hall, "Returning to Lukács: Honneth's Critical Reconstruction of Lukács's Concepts of Reification and Praxis," in *Georg Lukács Reconsidered. Critical Essays in Politics, Philosophy and Aesthetics*, ed. Michael J. Thompson (London/New York: Continuum, 2011), 197. In fact, following Feenberg, it is preferable to reconstruct the concept of second nature through the dialectic of structure and agency. Cf. Feenberg, "Lukács's Theory of Reification"; Feenberg, *Philosophy of Praxis*, 228–230.
75. Honneth, *Reification*, 22, 28.
76. Cf. ibid., 23.
77. See Feenberg's illuminating remarks on Lukács's non-reductionist theory of the relation between economy and culture, based on the concept of the form of objectivity in Feenberg, *Philosophy of Praxis*, 63–67.
78. Timothy Hall is right to note that "Honneth's engagement with Lukács's social theory is minimal and, in general, reliant on Habermas' critique in volume 1 of *The Theory of Communicative Action*"—therefore "he ends up imputing a base-superstructure model of society on Lukács and rejecting his account of the origin of reifying behavior as reductive." However, as Hall aptly stresses, "it is pretty clear that this is *not* what Lukács understands by the 'economic structure' of modern societies." Hall, "Returning to Lukács," 204.
79. Honneth, *Reification*, 28.

80. See Chari, "Political Critique," 598–601. Chari connects this inability with Honneth's "interactionist model of intersubjectivity," which "is less able to grasp the material conditions of social struggles" (ibid., 598). Other commentators make similar remarks on the limitations of a concept of reification that is too narrowly related to intersubjective relations. See e.g. Titus Stahl, "Verdinglichung als Pathologie zweiter Ordnung," *Deutsche Zeitschrift für Philosophie* 59/5 (2011): 737; Quadflieg, "Zur Dialektik," 707–708.
81. Cf. Lukács, *History*, 185–197.
82. Jütten formulates a similar critique of Honneth's anthropological orientation in "What is Reification?" 246–247.
83. Honneth, *Reification*, 55.
84. At least insofar as it is supposed to be understood as a *reinterpretation of Lukács*. Frederick Neuhouser rightly notes that only "very little of Lukács's original view–and even less of Marx's–remains in the theory of reification that Honneth develops." Frederick Neuhouser, "Axel Honneth: Verdinglichung." *Notre Dame. Philosophical Reviews*, March 7, 2006, https://ndpr.nd.edu/news/24979-verdinglichung/ (accessed 7 August 2015).
85. Feenberg critiques Honneth for implicitly succumbing to Habermas's dualism of system and lifeworld, though in his first book (Axel Honneth, *The Critique of Power. Reflective Stages in a Critical Social Theory*, Cambridge, MA: MIT Press, 1991) he had aptly criticized it. Cf. Feenberg, "Rethinking Reification," 108–109.
86. Lukács, *History*, 199.
87. See Kavoulakos, "Back to History?" 165–166.
88. See Feenberg, "Lukács's Theory of Reification"; see also Andrew Feenberg, *Alternative Modernity. The Technical Turn in Philosophy and Social Theory* (Berkeley/Los Angeles/London: University of California Press, 1995); *Questioning Technology* (London/New York: Routledge 1999); *Transforming technology. A Critical Theory Revisited* (Oxford/New York: Oxford University Press, 2002).
89. Also Jütten seems to agree with this claim. Jütten proposes a sympathetic reconstruction of the Lukácsian concept of reification through a charitable reading of Lukács's drawing upon German Idealism, which he interprets as an advocacy of "social freedom." Cf. Timo Jütten, "Andrew Feenberg, *The Philosophy of Praxis. Marx, Lukács and the Frankfurt School*," *Notre Dame Philosophical Reviews*, March 5, 2015, http://ndpr.nd.edu/news/56254-the-philosophy-of-praxis-marx-lukcs-and-the-frankfurt-school/ (accessed 30 July 2015). This reconstruction in some respects approaches Feenberg's interpretation. See Timo Jütten, "Verdinglichung und Freiheit," *Deutsche Zeitschrift für Philosophie* 59/5

(2011): 717–730. Also Stahl's suggestion that we understand reification as a "second order pathology" seems to update an essential part of the Lukácsian idea of a structural blocking of criticizing and changing the established social practices. Cf. Stahl, "Verdinglichung als Pathologie," 737–744.

90. Feenberg, "Rethinking Reification," 112.

Bibliography

Chari, Anita. "Toward a Political Critique of Reification. Lukács, Honneth and the Aims of Critical Theory." *Philosophy and Social Criticism* 36/5 (2010): 587–606.

Cutrone, Chris. "Andrew Feenberg, The Philosophy of Praxis: Marx, Lukács and the Frankfurt School"[review]. *Marx & Philosophy Review of Books*, February 14, 2015. http://marxandphilosophy.org.uk/reviewofbooks/reviews/2015/1524 (accessed 25 August 2015).

Dannemann, Rüdiger. *Das Prinzip Verdinglichung. Studie zur Philosophie Georg Lukács's*. Frankfurt/M.: Sendler Verlag, 1987.

Feenberg, Andrew. *Lukács, Marx and the Sources of Critical Theory*. Totowa, New Jersey: Rowman and Littlefield, 1981.

———. *Alternative Modernity. The Technical Turn in Philosophy and Social Theory*. Berkeley/Los Angeles/London: University of California Press, 1995.

———. *Questioning Technology*. London/New York: Routledge, 1999.

———. *Transforming Technology. A Critical Theory Revisited*. Oxford/New York: Oxford University Press, 2002.

———. "Reification and Its Critics." In *Georg Lukács Reconsidered. Critical Essays in Politics, Philosophy and Aesthetics*, edited by Michael J. Thompson, 172–194. London/New York: Continuum, 2011.

———. "Rethinking Reification." In *Georg Lukács: The Fundamental Dissonance of Existence*, edited by Timothy Bewes, and Timothy Hall, 101–120. London/New York: Continuum, 2011.

———. *The Philosophy of Praxis. Marx, Lukács, and the Frankfurt School*. London, New York: Verso, 2014.

———. "Lukács's Theory of Reification and Contemporary Social Movements." *Rethinking Marxism* 27/4 (2015): 490–507.

Hall, Timothy. "Returning to Lukács: Honneth's Critical Reconstruction of Lukács's Concepts of Reification and Praxis." In *Georg Lukács Reconsidered. Critical Essays in Politics, Philosophy and Aesthetics*, edited by Michael J. Thompson, 195–210. London/New York: Continuum, 2011.

Honneth, Axel. *The Critique of Power. Reflective Stages in a Critical Social Theory*. Cambridge, MA: MIT Press, 1991.

———. *Reification. A New Look at an Old Idea*, edited by Martin Jay. Oxford/New York: Oxford University Press, 2008.

Jaeggi, Rahel. "Verdinglichung – ein aktueller Begriff?" *Jahrbuch der Internationalen Georg-Lukács-Gesellschaft* 3 (1998/99): 68–72.

Jütten, Timo. "What is Reification? A Critique of Axel Honneth." *Inquiry: An Interdisciplinary Journal of Philosophy* 53/3 (2010): 235–256.

Jütten, Timo. "Verdinglichung und Freiheit." *Deutsche Zeitschrift für Philosophie* 59/5 (2011): 717–730.

———. "Andrew Feenberg, *The Philosophy of Praxis. Marx, Lukács and the Frankfurt School.*" *Notre Dame Philosophical Reviews*, March 5, 2015. http://ndpr.nd.edu/news/56254-the-philosophy-of-praxis-marx-lukcs-and-the-frankfurt-school/ (accessed 30 July 2015).

Kavoulakos, Konstantinos. "Back to History? Reinterpreting Lukács's Early Marxist Work in Light of the Antinomies of Contemporary Critical Theory." In *Georg Lukács Reconsidered. Critical Essays in Politics, Philosophy and Aesthetics.* edited by Michael J. Thompson, 151–171. London/New York: Continuum, 2011.

Lukács, Georg. *History and Class Consciousness. Studies in Marxist Dialectics*, trans. Rodney Livingstone. London: Merlin Press, 1971 (original work first published in 1923).

MacDonald, Agnes J.V. "Andrew Feenberg: *The Philosophy of Praxis: Marx, Lukács, and the Frankfurt School.*" *British Journal for the History of Philosophy* 22/4 (2014): 851–855.

Neuhouser, Frederick. "Axel Honneth: Verdinglichung." *Notre Dame. Philosophical Reviews*, March 7, 2006. https://ndpr.nd.edu/news/24979-verdinglichung/ (accessed 7 August 2015).

Quadflieg, Dirk. "Zur Dialektik von Verdinglichung und Freiheit. Von Lukács zu Honneth – und zurück zu Hegel." *Deutsche Zeitschrift für Philosophie* 59/5 (2011): 701–715.

Stahl, Titus. "Verdinglichung als Pathologie zweiter Ordnung." *Deutsche Zeitschrift für Philosophie* 59/5 (2011): 731–746.

———. "Verdinglichung und Herrschaft. Technikkritik als Kritik sozialer Praxis." In *Ding und Verdinglichung: Technik- und Sozialphilosophie nach Heidegger und der Kritischen Theorie,* edited by Hans Friesen, Christian Lotz, Jakob Meier, and Markus Wolf, 299–324. München: Wilhelm Fink, 2012.

CHAPTER 4

Gegenständlichkeit—From Marx to Lukács and Back Again

Christian Lotz

Introduction

Feenberg is one of the very few authors, particularly in the Anglo-American tradition, who has pointed out that the early development of critical theory cannot be understood without the inclusion of phenomenology and Neo-Kantianism. Although the connections between Adorno and phenomenology as well as between Marcuse and phenomenology are well known, the majority of contemporary debates in critical theory are rarely engaged with the rich historical context that provided the framework for early critical theory (in Germany), which was characterized by a tight intellectual network of phenomenological movements ranging from Husserl to Reinach, Neo-Kantian schools ranging from Windelband to Cassirer, philosophies of life, from Nietzsche to Simmel, as well as varying strands of positivism and early developments of analytic philosophy, such as the Vienna school. Although all of this is prominently featured in Adorno's inaugural lecture *The Actuality of Philosophy* (1931), which outlines the roadmap for early critical theory, recent critical school theorizing has lost sight of some, if

C. Lotz (✉)
Michigan State University, East Lansing, MI, USA

D.P. Arnold and A. Michel (eds.), *Critical Theory and the Thought of Andrew Feenberg*, DOI 10.1007/978-3-319-57897-2_4

not all, questions connected to this intellectual network. In addition, the "official" Habermasian and post-Habermasian Frankfurt School theorists, such as Honneth, Jaeggi, Benhabib, Allen, and Forst, no longer seem to be engaged in further developing social epistemology. Thus again, Feenberg is one of the few thinkers in critical theory who not only is aware of the rich historical horizon for a theory of society but who also has repeatedly pointed to the interdependency between epistemology and ontology, especially in connection with his Lukács interpretations.

As Feenberg rightly points out, one of the most important concepts in Lukács's theory is the term *Gegenständlichkeitsform* (objecthood, form of objectivity), which Lukács not only employs in order to translate epistemological concepts into social-ontological concepts but also uses in an attempt to transform the Kantian epistemological background into a social ontology within a Marxian framework.[1] To some extent, one could claim that Lukács is the "hinge" in early critical theory that holds the entire discourse together. However, in contrast to some interpreters, including Feenberg, I argue that the essays that form the first section of *History and Class Consciousness* are at least of equal importance to a proper understanding of Lukács, if not more important than the essay that most readers take to be Lukács's contribution, namely, the essay on reification. Accordingly, in my reading, I will shift the attention back to Lukács's methodologically grounded Marxism and extend Feenberg's intuitions about the concept of *Gegenständlichkeitsform* towards Marx's critique of political economy. I will do this by putting forward the view that Lukács's move from the "economic" concept of commodity to a cultural concept of commodity is already implied in Chap. 1 of *Capital*. As far as I can see, Feenberg almost exclusively focuses on the early Marx in his Marxist writings and dismisses the later writings. In contradistinction, I argue that we need to recover and renew the critical theory of society on the basis of a critique of political economy that is not misunderstood as a "labor theory of value," which, not incidentally, is a term that Marx never uses. It is in this sense that I am in agreement with more recent Marxist theorizing in Germany and Italy that is referred to as *value form theory*, the main proponents of which are Hans-Georg Backhaus, Helmut Reichelt, and Michael Heinrich. However, any dismissal of a social-economic theory of society has consequences for the critique of the capitalist social organization as a whole. As I have argued elsewhere, the loss of a theory of society and the turn of recent Frankfurt School thinkers to ethical and normative questions leads to the loss of a critical concept of capitalism, as this is no longer conceived as something that *can* be overcome. But critique for Marx, who follows Kant in this regard, means to demonstrate that the object in question, in Marx's case, capitalist society, is limited and finite.[2]

It seems to me that Feenberg's interpretation of Lukács's concept of *Gegenständlichkeitsform* not only needs to be tied back to Marx's *Capital* but also to a strong concept of categories such as that which characterizes a critique of political economy in the Marxian sense. A focus on categories is not only important for Feenberg and Lukács but also for all intellectual schools at the beginning of the twentieth century, insofar as the question of how to understand the concept of category is central for their attempts to escape metaphysics. Not only Husserl but also Lask and the early Heidegger are quite concerned with a reinterpretation of the concept of categories, insofar as these thinkers try to rescue it from what they conceive as its subjective background in Kant. For one, we find attempts to reinterpret categories as something that transcends the positioning of the transcendental subject, as something that is somehow given in life (Lask). For another, we find attempts to turn categories into units of meaning [*Sinn*] and "regional" frames (Husserl); and, finally, we also find the attempt to reinterpret categories through a hermeneutic lens, for example, in Heidegger, who approaches the problem of categories through his readings of Dons Scotus and Kant, transforming "categories" into what he calls in *Being and Time* "existentialia."

Moreover, further developing a theory of social categories is important for the following reasons: we should reject a transcendental theory of categories (Kant) that ties them to the structure of subjective reason; and, equally importantly, we should reject a metaphysical theory of categories (à la Hegel) that ties them to the structure of objective reason and to being. In my estimation, Marx's critical theory of society, long before phenomenology and Neo-Kantianism, tried to work out such a theory of categorization in reference to society, namely, as a *relation* that is constituted as something that exists *genuinely* between physical reality and subjective construction. This theory of the social as the "third" realm between subject and object as well as between mind and matter, in connection with the question of how we have access to the object "capitalist society," is the true core of Marx's *Capital*; and, as Lukács points out in his remarks on "orthodox Marxism" in *History and Class Consciousness*, removing this methodological aspect reduces Marxism either to sociology and empirical research or to a falsely understood naturalism.[3]

GEGENSTÄNDLICHKEITSFORM

Rather surprisingly for ahistorical readers of continental philosophy, it is not Heidegger but Emil Lask who speaks first of being as the "being of beings" in the sense of the "objectivity of objects" [*Gegenständlichkeit der Gegenstände*]. It is this formulation that turns transcendental philosophy into a new form of ontology insofar as the true question in transcendental philosophy and transcendental logic thus no longer simply asks about objects; instead, the question becomes what "makes" or constitutes objects as *these* and not other objects.[4] The underlying problem, then, is the question of whether there is some "super" form behind all types or regions of objects. As a consequence, the problem of the categorical constitution can be localized on two levels: on the one hand, categories constitute object regions (for example, one differentiates between animate and non-animate objects or between nature and history); and on the other hand, categories constitute objecthood or thinghood as such. The latter issue is clearly outlined by Kant in his *First Critique*; and for phenomenologists such as Husserl and Heidegger this becomes the problem of whether we can only develop a "formal ontology" or whether, as Heidegger claims, we can go further and develop a renewed fundamental ontology on the basis of explicating the history of metaphysics.

One might ask why all of this is of interest for social theory. We could respond by pointing to two major reasons that it is of interest: first, the concept of society could be conceived of as a region of being; and second, if social theory (in some sense) is supposed to be fundamental as the constitutive "realm" for all other regions (replacing Kantian reason and Husserlian pure consciousness), then it also has to be the "place" for constituting objects, as such. In the first case, we conceive of society as being based on a *specific* ontology. For example, we would try to separate "society" from other regions of being, such as nature or mind. In the second case, we conceive of society as a "super-region" that is in some form or another prior to the other regions of being. Lukács was indeed struggling with these two levels of social object constitution. This is especially clear if we trace the problems of *History and Class Consciousness* back to his earlier aesthetic writings. Further, as I will try to show later in my analysis, the attempt to bring both dimensions of categorical constitution and objecthood together is already implicit in Marx.

The transition to Lukács and the concept of *Gegenständlichkeit* can be seen in Lukács reflections on the concept of totality. According to Lukács, the problem of social object constitution is not only related to the question of how to turn the Kantian concept of objecthood into a social concept of "objectivity"; it is also *intrinsically* related to the concept of totality, as only the concept of totality is able to explain how the different "regions of being" can be conceived as having different forms of objectivity *while still* belonging to the *same Gegenständlichkeit*. However, this sameness, that is, its non-regional objecthood, should *not* be understood as *identity*, since understanding it as such would force us to move, again, towards an absolute dualism between form and content and an absolute distinction between regions and objects as such. These distinctions, in any case, need to be rejected from the standpoint of a Marxian-inspired social theory, since "society" is supposed to be the "region" *in* which the other regions are constituted. Accordingly, we should not conceive of "society" as something that is "above" or "beneath" all other categorically constituted realms. Instead, we need to understand the sameness of the regions as a *unity*; and it is precisely this move that Lukács is most concerned with in *History and Class Consciousness*. Now, for readers of Marx, this should not come as any surprise because in the *Grundrisse* Marx introduces the distinction between identity and unity in his remarks on methodological issues. We find the distinction addressed in three paragraphs. Here is the first:

> Hence this consumptive production – even though it is an immediate unity of production and consumption – is essentially different from production proper. The immediate unity in which production coincides with consumption and consumption with production leaves their immediate duality intact.[5]

Here is the second passage:

> The conclusion we reach is not that production, distribution, exchange and consumption are identical, but that they all form the members of a totality, distinctions within a unity. Production predominates not only over itself, in the antithetical definition of production, but over the other moments as well. The process always returns to production to begin anew.[6]

And this is the third decisive portion:

> The concrete is concrete because it is the concentration of many determinations, hence unity of the diverse. It appears in the process of thinking, therefore, as a process of concentration, as a result, not as a point of departure, even though it is the point of departure in reality and hence also the point of departure for intuition [*Anschauung*] and representation [*Vorstellung*].[7]

What Marx has in mind here—and Lukács, as an astute reader of these passages, must have seen it immediately—is the connection between epistemological and social categories *within* the horizon of a concept of identity that can no longer be conceived speculatively. This is to say that the concept of totality is here introduced to replace the concept of identity. As we are no longer dealing with a Hegelian speculative take on reality as the "self-moving concept," Marx returns to Kantian language by referring to "intuition" and "representation." However, Lukács must have been especially curious about Marx's claim in the last passage, which states that the point of departure for *both* intuition and representation is the concrete, that is, that which is supposed to be (re)presented by the concept of totality. One can easily see the relation to Lask's reference to the totality of life [*Lebenszusammenhang*] as the proper "place" for the emergence of categories, as well as to Lukács concept of history. However, in order to operate in a methodologically proper sense, Lukács argues that we need to remove the appearances from the "immediate form in which they are given" [*Gegebenheitsform*][8] and integrate and present these appearances within their unity. All of this is necessary for understanding social reality. Put differently, only if one understands each phenomenon that we isolate and analyze theoretically as being an effect of the whole is it possible to understand the singled out phenomenon as an *actual* phenomenon. If we were to remain at the level of positivities, each isolated phenomenon would remain abstract. As we know, Marx tries to solve this problem of unity and diversification by adopting a *genetic* approach to social categories. For this part, Lukács introduces the concept of *Gegenständlichkeitsform* via a critique of the concept of causal reciprocity [*Wechselwirkung*]:

> It must go further in its relation to the whole: for this relation determines the objective form of every object of cognition. Every substantial

> change that is of concern to knowledge manifests itself as a change in relation to the whole and through this as a change in the form of objectivity [*Gegenständlichkeitsform*] itself. Marx has formulated this idea in countless places. I shall cite only one of the best-known passages: 'A negro is a negro. He only becomes a slave in certain circumstances. A cotton-spinning jenny is a machine for spinning cotton. Only in certain circumstances does it become capital. Torn from those circumstances it is no more capital than gold is money or sugar the price of sugar.' Thus the objective forms of all social phenomena change constantly in the course of their ceaseless dialectical interactions with each other. The intelligibility of objects develops in proportion as we grasp their function in the totality to which they belong. This is why only the dialectical conception of totality can enable us to understand *reality as a social process*.[9]

We are now able to see the connection that Lukács establishes between the form of objects to be found in "regions" of society and objecthood as such. Both dimensions are tied to one process, that is, the historical reality, of which they are both parts. As we know from Feenberg's illuminating remarks, the unifying form that Lukács has in mind is a cultural form that he identifies with reification, one based on Lukács's (implicit) "transcendental account of meaning,"[10] insofar as "meaning is the 'being' of the phenomena through which we gain access to them as what they 'are.'"[11]

What Is a Category?

The problem of categories is tied to the history of philosophy; and, in the confines of this paper, I am neither able to develop a satisfactory general account of it nor outline further steps toward a more satisfactory theory of social categories.[12] Nevertheless, two consequences of the foregoing remarks can be identified: (1) The task is to develop a theory of social categories that no longer treats the social as a "sub-domain" of reality; instead, one needs to argue from a Marxian perspective that the social domain is in some sense the *network of relations itself*, that is, that it is nothing other than the genetic relations between social categories (if we assume that categories are not "things," but are *relations*) which, in turn, necessarily involves the idea that society is a quasi-transcendental form. It seems to me that Lukács (rightly) never gave up on this idea. (2) We should not give up the task of developing a social epistemology, which was called for by Habermas in his *Theory of Communicative*

Action; rather, we need to follow the insights of Lukács further and ask in which sense ontology and epistemology can be brought together in *one* theory. Though I am unable to present the solution in this paper and I have doubts that Lukács's subject–object differentiation will solve the problem, it seems to me that the problem of categories in Marxian thought should be rethought from a phenomenological perspective, insofar as a phenomenological approach can avoid the pitfalls of both Kantian subjectivism and Hegelian objectivism, given that in phenomenology "the categories can be 'read off' for the first time in their *full content*."[13] Heidegger's early reflections on categories in his *Habilitationsschrift* can function as a guiding thread for this broader task. In a telling reflection on his Dons Scotus book, Heidegger introduces the problem of categories as *both* an epistemological *and* ontological task, first introducing categories in the sense I had in mind above in the following way:

> If we conceive of the categories as elements and means for interpreting the sense of what is experienceable—of what is an object in any sense—then what ensues as a basic requirement for a theory of categories is *characterizing and demarcating the different domains of objects into spheres that are categorially irreducible to one another.*[14]

Here Heidegger deals with the concept of categories as a regional concept and a concept of diversification before he introduces it in a second, more important sense:

> A category is the most general determination of objects. Objects and objectivity *[Gegenständlichkeit]* have, as such, meaning [Sinn] only *for* a subject. In this subject, objectivity *[Objektivität]* is built up through judgments. Consequently, if we want to conceive of categories in a decisive manner as determinations *of objects,* then we must establish their essential relations to the forms that build up objectivity *[Gegenständlichkeit].*"[15]

It is clear that Heidegger is still struggling with the problem of Kantian subjectivity here, as a relation to what a thing is cannot be found *in* the object and needs to be understood as its transcending condition. To borrow a formulation by Steven Crowell, categories open up the "possibility of the prejudicative open availability of objects."[16] However, already at this point, we can see how the same problem returns when we apply it to the problem of society and social constitution insofar as the problem is not that we encounter social *objects*. Instead, what is astonishing is that we do not simply encounter objects; instead, we are encountering *social* objects. As a consequence, the problem of *Gegenständlichkeitsform* returns with full force, and it is clear how we should address this distinction in Marxist theory. In Marxist theory, we need to argue that the objects are not simply encountered as social objects; rather, they are encountered as social objects of a specific, historically limited form that is at present, in their capitalist social form.[17] Only the acknowledgment of such a historically limited form will help us overcome a universal concept of history, which tends to fall back on naturalism or a quasi-natural teleology.

Feenberg's Misreading of Economic Categories

Despite Feenberg's absolutely crucial reading of Lukács and the concept of *Gegenständlichkeit*, I wonder whether his tendency to dismiss political economy leads him to an understanding of social categories that violate the principles introduced by his own affirmation of the quasi-transcendental concept of social form. In one of his central essays on reification he writes the following:

> Reification is in the first instance practical rather than theoretical. That is to say, the reification of social reality arises from the way individuals act when they understand their relation to social reality to be reified. The circularity of reification is a familiar social ontological principle currently referred to by the fashionable term "performativity." For example, money is money only insofar as we act as though it were money and it is the success of this sort of action that determines our conviction that money is in fact money. In behaving as friends we constitute a relationship which we perceive as a substantive thing, a friendship. A tool is only a tool insofar as we perceive *its* toolness in the possibility of a specific type of use. Social "things" are not merely things but are implicated in practices. The categories under which social life makes sense are the categories under which it is lived.[18]

Though I tend to agree with Feenberg's focus on praxis, I think that we need to reject the strong constructivism that moves into his understanding of social categories such as money. We can do this on the basis of Lukács, though, insofar as he claims that social categories, such as money, are not "separated" things that receive their meaning through "actions" and that, instead, each social category can only receive its meaning through the genetic relations that are contained in them and are *unified*. Put differently, though Feenberg is right in pointing to social praxis as the underlying "essence" of social things, such as technologies, we need more carefully argue in regard to social *categories*, since categories are *explicitly* constituting the *social form as a whole*. Social categories, then, can only be changed when the whole changes and, as a consequence, a theory of revolution or radical change must be combined with an epistemology of the categories themselves. In the case of money, this is even more important than in other examples, as it is clear that the entire problem of money can be disconnected neither from the commodity form nor from the capital form. As Marx argues, money in the capitalist social organization can *only* be understood as a form of capital. It is already determined by the capital form in each of its movements, and it can, therefore, be related to the unity, that is, the totality of capitalist social organization. As a consequence, the meaning of each category can only change if all relations that are genetically contained in each category change. However, Feenberg's interpretation of money in the above quote comes close to a subjective concept of money (and value) because it seems to imply that we could simply stop taking money as money. I reject this position, however, since the "meaning" of money can only change if we would live in a reality in which money would no longer genetically imply capital form and commodity form, which, in turn, presupposes that the unity (that is, totality) of these relations has shifted.[19] Put in phenomenological language, we could also say that the *world* in which money exists as this specific, for example, capitalist form of money, must change. In order to clarify this further, let me now turn to a re-reading of Chap. 1 of *Capital*, since two crucial distinctions that Marx introduces at the beginning of *Capital*, namely, the distinction between commodity and commodity *form*, as well as the distinction between use value and exchange value, are essential for a proper understanding of his thought. Indeed, the latter distinction is especially important in regard to Marx's claim that value as the constitutive unity of capitalist society is neither a physical nor a mental phenomenon.

A Re-reading of Chapter One of *Capital* and the Return to Marx's Critique of Political Economy

The distinction between use value and exchange value has been discussed often in the literature, and I do not intend to repeat the most obvious views about that here. However, it is of crucial importance to note that this distinction tends to be treated as an absolute distinction, that is, as if we can encounter things in capitalism as either use values or exchange values and as if in non-capitalist societies there would simply be use values. For example, David Harvey's most recent book, unfortunately, presents a reading that artificially separates use and exchange value, identifying them as two ways in which commodities can function in capitalism.[20] As I argue in the following, this reading is misguided and misses the central point that Marx works out in chapter one of *Capital*, namely, that things in a society that finds its unity in money and is actualized as abstract labor can *never* be "just" use values, that is, things to be used. Three main reasons for this thesis are the following.

First, in the further development of the categories in *Capital*, the distinction between use value and exchange value *becomes* an internal distinction of capital and abstract labor. Accordingly, although Marx *seems* to introduce the distinction as an absolute distinction between use value as something that exists in all social formations, and exchange value that only has a place in particular societies that are determined by the capitalist mode of production, the further discussion of the internal relations implied in categories such as money and capital reveals that use value is a dialectical concept. As a consequence, the concept of use value should not be conceived as something "external" or outside the system. Use value, in other words, takes on a *specific* form under which it exists in capitalist social formation. Only this allows Marx to speak of a contradiction between use value and exchange value. If the use value of commodities were really independent of its *being* a commodity, then a contradiction *in* capitalism *between* use and exchange could never occur, since one side of the contradiction would not belong to capitalism *specifically*. In addition, in the nineteeth century, "Gebrauchswert" was used as with a synonym for "Gebrauchsding," that is, use thing or useful thing. To be sure, the usage of the term "value" is confusing, as it seems to indicate that Marx refers to a thing and its use value as something that a thing possesses as a property. However, a close reading of the text should allow readers to come to the conclusion that by "use value" Marx refers to the *thing*. For example, in his important clarifications that Marx offers in 1881 under the title *Notes on Adolph Wagner*, he writes the following:

> What I proceed from is the simplest social form in which the product of labour presents itself in contemporary society, and this is the "*commodity*." This I analyse, initially in the *form in which it appears*. Here I find that on the one hand in its natural form it is a *thing for use [Gebrauchsding]*, alias a *use-value*; on the other hand, a *bearer of exchange-value*, and from this point of view it is itself an "exchange-value." Further analysis of the latter shows me that exchange-value is merely a "*form* of appearance," an independent way of presenting the *value* contained in the commodity, and then I start on the analysis of the latter.

In these notes, Marx not only repeatedly points out that by "use value" he means "natural thing," but also criticizes Wagner for constantly confusing "use value" and "value" because the words (unfortunately) indicate that both refer to entities that possess "values" (*Gebrauchswert* vs. *Tauschwert*). However, as Marx underlines in these notes, the *value* in the sense in which he introduces the term in *Capital* should be distinguished from both use value and exchange value.[21]

Second, the distinction between use value and exchange value is primarily not introduced in order to indicate two different ways in which commodities "function" in our economy; rather, Marx connects the distinction between use and exchange to his thesis that value as the *unity* of capitalist social organization should not be conceived as something *natural*. The main line of argument for the purely social nature of value as the unity of capitalist society and the form of all relations is the following: the usage of a thing is determined by its natural properties. For example, whether sand or pizza can satisfy our hunger ultimately depends upon its natural properties or, if we also take intellectual use things into account, its thing properties, that is, properties that it has as this or that thing. Moreover, the properties that a thing possesses as a thing are *non-relational*, that is, they do *not* exist in relation to something other than themselves. Now, if we reconstruct how Marx proceeds in Chap. 1 of *Capital*, it seems at first as if he wants to say that *one* thing has a use value and an exchange value. However, a more careful reading reveals that this assumption is incorrect. In fact, Marx himself states this later after he introduces the distinction:

> When, at the beginning of this chapter, we said in the customary manner that a commodity is both a use value and an exchange value, this was, strictly speaking, wrong. [...], and the commodity never has this form when looked at in isolation, but only when it is in a value-relation or an exchange-relation.[22]

Accordingly, what he says is that one thing functions as a use value, that is, a thing, and another thing functions as an exchange value, that is, the other thing. Put differently, it is *never* the case that one thing "has" or possesses an exchange value; rather, it *is* the exchange thing for another thing. Consequently, the exchange value of a thing is not somewhere to be found "on" or "in" it; instead, the *other* thing is the exchange value of the first thing. This crucial move is decisive, insofar as Marx states from the beginning that exchange value and use value only exist in and as a relation, and never in themselves. This is also the reason for the fetishism connected to prices, insofar as price tags that things carry around like name tags hide their relationality. Their relationality gets lost and appears as what they are not, namely, natural properties, that is, properties of the thing itself. Again, only if we understand the relationality that Marx introduces at the beginning of *Capital* can we also understand why value is the condition for the possibility of that relationality. Indeed, the *exchangeability* itself can then be introduced by Marx as a social concept. As stated in *Capital*, value is the "form of direct exchangeability,"[23] which is to say that the exchangeability appears to us as something immediate and something that we do not need to establish, as it is presupposed as the *Gegenständlichkeitsform*. A commodity "*seems* to be endowed with its equivalent form, its property of direct exchangeability, by nature" (emphasis, C.L.).[24] However, the equivalent *form* is "purely social."[25] Value is the expression of the unity of all social relations and never anything that can be found in a thing. If value could be found in any other property in a thing, then we would fall back to a naturalist theory of value in which value magically emerges out of the use of commodities. However, if Marx is correct with his focus on the relation *between* things, then all value theories that begin with utility and the demand for commodities are non-starters, as they do not understand that economic theory is a theory of society and not a theory of things that people exchange because of their internal properties.

> Not an atom of matter enters into the objectivity of commodities as values; in this it is the direct opposite of the coarsely sensuous objectivity of commodities as physical objects; […] their character as values is therefore purely social.[26]

Moreover, he says the following:

> This common element [value, C.L.] cannot be a geometrical, physical, chemical or other natural property of commodities.[27]

And, finally, he concludes:

> So far, no chemist has ever discovered exchange-value either in a pearl or a diamond. The economists who have discovered this chemical substance, and who lay special claim to critical acumen, nevertheless find that the use-value of material objects belongs to them independently of their material properties, while their value, on the other hand, forms a part of them as objects.[28]

Put differently, that which constitutes commodities as commodities, that is, their form, cannot lie in their "thing properties" alone. Accordingly, it is incorrect to claim that commodities can function as use values once they leave economic transactions.

Marx's *Capital* is from the beginning a rejection of any utility maximization theory, as he claims that capitalist social organization is *totally* independent of how individuals weigh and compare the "value" of things they buy and sell. Instead, he tries to show that we depend upon an objective structure that makes all things "on the ground" possible. As a consequence, the *reality* of capitalist social organization is not constituted by desires and wants; rather, it is precisely the other way around: desires, wants and needs are produced by them, being subjected to the valorization process and subjected to abstract labor as the intellectual and bodily totality of human capacities, which include our noetic apparatuses. With the introduction of the wage into the social system, it becomes possible to monetize the *entire social order*, as labor power is (1) to be found universally in every individual on earth (whereas concrete labor is diverse) and (2) money can itself become universal.

Third, Marx introduces a crucial distinction between commodity and commodity form, which is already implied in my two foregoing points.

Marx is not interested in reflecting on commodities as "things" that have properties independent from their social form; and, as a consequence, we can no longer argue that commodities, once they are taken out of their economic circulation, are no longer commodities. For example, one could argue that the pizza that you buy for a dinner with friends is a commodity in the store and during the transactions connected to it, but that it is no longer functioning as a commodity when you enjoy it with your friends. However, on the basis of what Marx develops at the beginning of *Capital*, we need to reject this way of understanding commodities, since Marx is not really interested in commodities as "beings." He is reflecting on the being of beings, or, to use words from before, he introduces the commodity *form* as *precisely* that which Lukács calls *Gegenständlichkeitsform* (the form of how objects are encountered in capitalism, that is, their objecthood). The commodity form is that which constitutes commodities *as* commodities, and it is only because of this social form that things that are not yet commodified can *become* commodified and can take on this specific social form. Accordingly, your dinner is all the way through "commodified," especially if we take into account that most likely everything you "use" for that dinner was at some point produced and bought under capitalist conditions. Again: if commodities were commodities because of some natural properties that belonged to things like their weight and material, then the social form, as the condition for the possibility of it being a thing in the capitalist social organization, would be natural, that is, it would *not* constitute its historical specificity. Value, as Marx puts it, is the "*general* social form" (CI, 159; emphasis, C.L.), which means that potentially everything is subjected to this form, and, as a consequence, the commodity form is all-pervasive. What Marx tries to tackle in Chap. 1 (and then in Chap. 5 on labor power) is an explanation for why the commodity form as a *cultural* form and as the horizon of meaning in capitalist society can be *universal*, that is, he tries to explain how it can be the general form under which everything in capitalism is constituted.

As I have argued elsewhere, it is therefore totally misguided to understand Marx merely as a theorist of a "labor theory of value" or merely as a theorist who is simply concerned with the "capitalist mode of production," insofar as this mode is from the beginning conceived as one in which value is the *general social form*. Consequently, value is the all-pervasive form of *all* relations that constitute the unity and integration of *all* social agents and *all* things under one really abstract form.[29] We should, therefore, be cautious about talking about "commodities" as things that are around us. Things around us are characterized by their internal properties that determine what they can be used for, but it is precisely Marx's argument that their exchangeability does not depend upon internal properties; rather, the content of their exchangeability as their *Gegenständlichkeitsform* is a social reality. The capitalist social organization, as we should conclude from Chap. 1 of *Capital*, can neither be naturally derived, which means that teleological approaches to social development are excluded, nor be a result of mental constructions and "beliefs" that might or might not underlie human relations. Put differently, the necessity of social unity neither stems from the arbitrary "encounter" of individuals nor does it issue from some hidden meta-historical process that explains why we "naturally" ended up in capitalism. *Capital*, strictly speaking, is an anti-teleological book and this is the reason for why we should not refer to it as exemplifying "Historical Materialism."

Consequences: Feenberg, Technology, Political Economy

In sum, I have argued that Lukács's concept of *Gegenständlichkeitsform* as the "unifying pattern of an entire society"[30] (culture) is already implied in the Chap. 1 of *Capital*, and that Lukács's genius was to develop his theory without having the entire Marx scholarship available, which is what my contemporary reading of Marx is based on. Marx's claim that "the commodity form is the *universal* form of the product of labor" (emphasis, C.L.)[31] is, in its roots, already the claim that the commodity form is, as Feenberg puts it in relation to Lukács, a *culturally universal* form. As Feenberg has it in regard to reification, "reification is the underlying unity of the social system, the 'model' of 'all the forms of objectivity of bourgeois society together with all the corresponding forms of subjectivity'."[32] Though this seems to be a philological point, I submit that a re-reading of Marx on the grounds presented in this essay should shift our focus back to the critique of political economy as the fundament for the further development of critical theory.[33] It should also be helpful for the further development of a critical theory of society that takes epistemological issues *as* ontological issues seriously.

Luckily, due to his focus on technology as the mediating concept, Feenberg's approach to critical theory offers us major clues for further conceptualizations and stands out from discussions that are focused exclusively on power, identity, and normativity alone.

Notes

1. I should underline that the following reflections go back to a conversation that I had with Andrew Feenberg in 2015 at the conference of the *Society for Phenomenology and Existential Philosophy (SPEP)*. Feenberg's lucid explanations of Lukács's reflections on *Gegenständlichkeit* greatly inspired the following essay, especially since I had not seen the connection that my own attempt to translate Kantian concepts into social concepts shares with Feenberg's lucid interpretation of Lukács's concepts. For this, see Christian Lotz, *The Capitalist Schema. Time, Money, and the Culture of Abstraction* (Lanham: Lexington Books, 2014).
2. For this, see Christian Lotz, "An der Oberfläche der Tauschgesellschaft. Kritik der Kritischen Theorie," *Prokla. Zeitschrift für kritische Sozialwissenschaft*, 180, 2015, 453–469.
3. György Lukács, *Geschichte und Klassenbewusstsein*, Werke, Frühschriften, Band II (Bielefeld: Aistheis, 2013), 175.
4. For this, see Lask's letter to Rickert in Emil Lask, *Die Logik der Philosophie und die Kategorienlehre. Eine Studie über den Herrschaftsbereich der logischen Form* (Siebeck: Tübingen, 1911/1993), 272.
5. Karl Marx, *Grundrisse*, tr. Martin Nicolaus (London: Penguin, 1993), 91.
6. Marx, *Grundrisse*, 102.
7. Marx, ibid., 101. I altered the translation, as the English translator of the *Grundrisse* unfortunately translates "Anschauung" with "observation" and "Vorstellung" with "conception," which hides the fact that Marx obviously operates with Kantian terms here.
8. Lukács, *Geschichte*, 179 (translation altered).
9. Lukács, ibid., 13.
10. Andrew Feenberg, *The Philosophy of Praxis: Marx, Lukács, and the Frankfurt School* (London: Verso 2014), 66 and 73. Chapter four of the *Philosophy of Praxis* is one of the richest accounts of these problems that I have ever read, as Feenberg delivers in it an ingenious and crystal-clear account of the underlying philosophical concepts in Lukács that are related to the commodity form.
11. Ibid., 75.
12. I am working on an extensive project tentatively entitled *Phenomenology of Capital* in which I intend to present a core theory of social categories as a theory of social reality.
13. Martin Heidegger, *Supplements: From the Earliest Essays to Being and Time and Beyond*, ed. John van Buren (New York: SUNY Press, 2002), 65.

14. Ibid., 63.
15. Ibid., 64 [revised translation].
16. Steven Crowell, *Husserl, Heidegger, and the Space of Meaning. Paths Toward Transcendental Phenomenology* (Evanston: Northwestern University Press, 2001), *53*.
17. For this, also see Feenberg, *Philosophy of Praxis*, 88.
18. Andrew Feenberg, "Reification and its Critics," in "*György Lukács Reconsidered: Essays on Politics, Philosophy and Aesthetics*," ed. Michael M. Thompson (London: Routledge, 2011), 172–194, here 179.
19. On a side note, Lukács himself overlooks the crucial importance of money for the entire critique of political economy. For a massive critique of Lukács's failure to properly understand money see chapter two in Frank Engster, *Das Geld als Mass, Mittel und Methode. Das Rechnen mit der Identität der Zeit* (Berlin: Neofelis Verlag, 2014).
20. For this, see David Harvey, *Seventeen Contradictions and the End of Capitalism* (Oxford: Oxford University Press, 2014), chapter one.
21. Karl Marx, *Marx's Notes on Adolph Wagner German and English*, online: https://www.marxists.org/archive/marx/works/1881/01/wagner.htm [last accessed July 30, 2016].
22. Karl Marx, *Capital, Volume I*, tr. Ben Fowkes (London: Penguin 1990), 152. This comes close to how Feenberg puts it: "Particular social objects [...] cannot be understood in isolation, but only in relation to the whole because that relation is constituting for their meaning" (Feenberg, *Philosophy of Praxis*, 76).
23. Ibid., 154.
24. Ibid., 149.
25. Ibid., 139.
26. Ibid., 138.
27. Ibid., 127.
28. Ibid., 177.
29. For this, see Lotz, *The Capitalist Schema,* chapter one + two.
30. Feenberg, *Philosophy of Praxis*, 64.
31. Marx, *Capital*, 152.
32. Feenberg, *Philosophy of Praxis*, 66.
33. Beside the German value form theorists, the research network around Werner Bonefeld is also working on a renewed critical theory with focus on political economy. For this, see Werner Bonefeld, *Critical Theory and the Critique of Political Economy: On Subversion and Negative Reason* (London: Bloomsbury Press, 2014), and my comments on Bonefeld in Christian Lotz, "Critical Theory and the Critique of Political Economy: On Subversion and Negative Reason," *Radical Philosophy Review*, 18/2, 2015, 337–342.

Bibliography

Bonefeld, Werner. *Critical Theory and the Critique of Political Economy: On Subversion and Negative Reason*. London: Bloomsbury Press, 2014.

Engster, Frank. *Das Geld als Maß, Mittel und Methode. Das Rechnen mit der Identität der Zeit*. Berlin: Neofelis Verlag, 2014.

Feenberg, Andrew. "Reification and its Critics." In "*György Lukács Reconsidered: Essays on Politics, Philosophy and Aesthetics*." Edited by Michael M. Thompson. London: Routledge, 2011, 172–194.

———. *The Philosophy of Praxis: Marx, Lukács, and the Frankfurt School*. London: Verso, 2014.

Harvey, David. *Seventeen Contradictions and the End of Capitalism*. Oxford: Oxford University Press, 2014.

Heidegger, Martin. *Supplements: From the Earliest Essays to Being and Time and Beyond*. Edited by John van Buren. New York: SUNY Press, 2002.

Lask, Emil. *Die Logik der Philosophie und die Kategorienlehre. Eine Studie über den Herrschaftsbereich der logischen Form*. Siebeck: Tübingen, 1911/1993.

Lotz, Christian. *The Capitalist Schema. Time, Money, and the Culture of Abstraction*. Lanham: Lexington Books, 2014.

———. "An der Oberfläche der Tauschgesellschaft. Kritik der Kritischen Theorie," *Prokla. Zeitschrift für kritische Sozialwissenschaft*, 180, 2015, 453–469.

——— "Critical Theory and the Critique of Political Economy: On Subversion and Negative Reason." *Radical Philosophy Review*, 18/2, 2015, 337–342.

Lukács, György. *Geschichte und Klassenbewusstsein*, Werke, Frühschriften, Band II. Bielefeld: Aistheis, 2013.

Marx, Karl. *Capital, Volume I*. Translated by Ben Fowkes. London: Penguin, 1990.

———. *Grundrisse*. Translated by Martin Nicolaus. London: Penguin, 1993.

CHAPTER 5

Feenberg, Rationality and Isolation

Clive Lawson

An important theme within critical theory has been the extent to which social systems appear to be "rational." An interesting manifestation of this is to be found in the re-emergence of interest in the concept reification. This concern with reification is linked to attempts to explain why capitalism not only survives but appears as the best or only possible organising system for human societies. At the heart of the latter endeavour is the idea that various social phenomena, such as the working of markets, the organisation of institutions, and the emergence and adoption of new technology, all appear as in some sense *rational*, so making attempts at criticism or challenge appear as misguided or even regressive.

Perhaps the most prominent recent formulation of these ideas is to be found in the work of Andrew Feenberg, who both accepts the critical theory problematic and develops it in interesting and innovative ways. In this chapter, I shall be concerned with the conception of rationality at work in Feenberg's contributions and its relationship to science. I shall then turn to a conception of science that is usually associated with the work of critical realists. The aspect of this conception that I focus upon is the way in which the different practices of science depend upon

C. Lawson (✉)
Girton College, University of Cambridge, UK

D.P. Arnold and A. Michel (eds.), *Critical Theory and the Thought of Andrew Feenberg*, DOI 10.1007/978-3-319-57897-2_5

(or are a reaction to) the different degrees of isolatability of different causal mechanisms to be found in different domains. This account is then drawn upon to provide an alternative grounding for the conception of rationality found in Feenberg's contributions. First, however, I shall contextualise this discussion by giving a very brief account of the emergence of rationality as a central concept in critical theory, and its relationship to ideas of reification.

Reification: Marx to Marcuse

The starting point of critical theory's focus upon ideas of reification is usually located in the work of György Lukács.[1] Lukács, drawing explicitly upon Marx's ideas about commodity fetishism, argues that within capitalism social relationships between people take on a different quality:

> The essence of the commodity-structure ... is that a relation between people takes on the character of a thing and thus acquires a 'phantom objectivity', an autonomy that seems so strictly rational and all-embracing as to conceal every trace of its fundamental nature: the relation between people.[2]

Notice that although rationality is mentioned, the operative term here is actually autonomy. A relation between people is experienced as if it is in fact quite independent, or autonomous, of people. For example, the market can be explained in terms of the fetishism of commodities, where there is a substitution of exchange relations for human relations. This projects the idea of a "law" of markets which takes on an independent existence. Such phenomena as markets and prices have a curious existence in Lukács's account: they are not imaginary, but neither are they in reality fixed or given independently of the actions of people; rather they are *reified*. Part of the process of challenging the prevailing state of affairs, Lukács argues, involves a process of coming to understand social relations, especially their contingency and fragility. In so doing, social structures become de-reified.

For Lukács, the manner in which capitalism separates workers from the means of production, and then manages to present the resulting relations as outside their control, serves as a model for understanding the whole of society. In developing this idea, Lukács accommodates ideas often thought at odds with (or even an alternative to) Marx's basic account. The most important example of this is Max Weber's idea of rationalisation.[3] For Weber, specialisation and the division of labour lead to an institutional complexity that is difficult to control. In response, various strategies emerge that involve the quantification or quantitative assessment of all kinds of activities, especially work that can thus be evaluated by bureaucrats at a distance, with little knowledge of the processes in question. This means that bureaucracies become more significant (and powerful) as the population and complexity of their societies grow. Such rationalisation, moreover, both gives rise to a growing mindset of calculation and control, and encourages a view of organisational forms as increasingly beyond dispute and rational.[4]

In taking on board these Weberian ideas, Lukács is suggesting that the commodity fetishism stressed by Marx is only a special case of the problems of capitalism. There are more general cultural problems, and a series of phenomena that contribute to normalising the formalistic, quantifying tendencies in social life. In other words, Lukács's use of the term reification is intended to combine both Marx's fetishism and Weber's rationalisation.[5] In combining these ideas, however, he tends to merge the different meanings of "reified" as referring to social processes that are either falsely perceived as autonomous (and so unchangeable) or as rational (and so may be changeable, but not for the better).

Later critical theorists, in particular the Frankfurt School of critical theory, emphasised the importance of the role played by rationality in Lukács's account.[6] At least part of the reason for this emphasis is that these theorists move further away than Lukács from the central role played by commodity fetishism in Marx's account. In particular, Adorno and Horkheimer de-emphasise the central role given by Lukács to the working classes, both in terms of their experience of capitalism and their position as the prime agents of change. Instead, Frankfurt School theorists focus upon the extension of reification into all social life, in particular the media, propaganda, and consumerism that become central to all aspects of everyday life and consciousness. But in so doing, the rationality of these forms becomes the main focus.

For the most part, Frankfurt School theorists tend to adopt a very negative or dystopian outlook, where everyday experience within capitalist society is radically distorted by the commodification of just about everything. One exception to this tendency is the more positive contribution of Marcuse.[7] Marcuse was the first of these writers to be positive about technology. In large part, this positive orientation is an outcome of the idea that radical change must involve not only social transformation but also technological transformation. Marcuse accepts much of the analysis of capitalism put forward by other Frankfurt School theorists, including their focus upon rationality and its relationship to science. His reaction to such problems, however, is to direct blame at the conception of science that has come to dominate social life.[8]

From Reification to Instrumentalisation

The currently most prominent, and I believe most developmentally consistent, treatment of these ideas is provided by Feenberg.[9] Feenberg, a student of Marcuse, carries on the line of critical theory developments pretty much where Marcuse left off. In particular, his explicit focus is upon a wide range of institutional processes within capitalism that appear, or are experienced, as rational and (thus) beyond challenge. But Feenberg, although otherwise generally favourably disposed to Marcuse, is critical of his calls for a whole new science. Indeed, Feenberg is very explicit in stating that his own critical theory is a theory of technology, not a theory of science. As with most other later critical theorists, Feenberg also de-emphasises what he sees as Lukacs's (and Marx's) preoccupation with the change that might result from working class resistance, and the privileging of the industrial experience of the working classes, emphasising that a whole host of modern institutions take on this reified form and are just as significant for both understanding capitalism and changing it.

A crucial point, for Feenberg, is the idea that a better society, such as socialism, would not depend for its own existence upon the systematic misrecognition or misrepresentation of itself (whether this be as fixed, natural, or optimal). Feenberg draws attention to conflicts that have emerged in reaction to institutional excesses (environmentalism, medical pressure groups, etc.), which serve both to enlarge the public sphere and anticipate democratic forms of society. According to Feenberg, specialisation and control, which are intrinsic to institutions, always generate resistance, so that the goal must be to formalise such resistance and expand democratic input at every stage.[10]

In arguing his case, Feenberg focuses upon technology as an example that can be generalised to other (reified) institutions. Central here is a distinction drawn between two levels of what he terms instrumentalisation. The first, or primary, instrumentalisation loosely corresponds to the technical orientation towards reality identified by Heidegger as the technological "mode of revealing."[11] The underlying process is one in which objects are isolated and exposed to external manipulation. However, for such objects to be realised as working technology, to take on a real form in actually existing devices systems and networks, a secondary instrumentalisation is required that emphasises the social, meaningful nature of technology. Strategically, this move serves to combine phenomenological insights from Heidegger, Marcuse, Habermas and others with constructivist insights concerning the irreducibly social constitution of technology.

Feenberg argues that technical activity is always dialectical in the sense that it always involves these two levels of instrumentalisation, and thus moments of separation and of integration.[12] This is important in Feenberg's account, as it is the latter moment of integration that compensates for capitalism's reifications (and so rectifies the especially negative evaluation of technology amongst much of critical theory). Although these two moments are always inextricably linked, the relation between them is open to significant historically contingent differences. Under capitalism, there is a tendency to reduce technical activity to the primary level, of de-contextualisation, calculation, and control. Secondary instrumentalisation tends to be considered as non-technical and so is suppressed or devalued.[13] This is particularly so in the case of the labour process, where obstacles to secondary instrumentalisation are to be found wherever integrative technical change would threaten control. In particular, the integration of skill and intelligence into production is often resisted due to the fear that the firm will become dependent upon its workforce.

More specifically, Feenberg suggests that there are several reifying moments of technical activity, each with its own compensating integrative moment that is severely restricted under capitalism.[14] Any better society must find ways of more fully integrating these moments without restricting the secondary instrumentalisations in the way that capitalism does. Feenberg suggests that today secondary instrumentalisations take the form of afterthoughts, usually after protest.[15] A better society would make them structurally part of the same process.

To achieve a better society, Feenberg argues, we do not require some sophisticated central plan for technology, however, but what he terms genuine technological progress. To clarify what he means by this, Feenberg draws upon the work of Gilbert Simondon.[16] On Simondon's account, technological progress is seen as a process of discovery of synergisms between technologies and their various environments. Simondon situates technologies along a continuum from less to more integrated designs. In loose designs each part performs a separate function; in the course of technological development parts are redesigned to perform many functions and structural interactions take on many functional roles—thus becoming more *concrete*. For example, an engine combines surfaces used to dissipate heat merged with those used to reinforce the engine case (that is, two structures with separate functions are combined within a single structure with two functions).

Once they have evolved into complex mechanisms, technologies also in a sense generate the environmental conditions to which they adapt. This involves constantly incorporating and internalising features of the environment in which the technical object is to operate; for example, cars become more comfortable whilst reducing emissions and including safety features. In this way, Feenberg suggests, concrete technology "includes nature in its very structure … [and this] passage from abstract technical beginnings to concrete outcomes is a general integrative tendency of technological development that overcomes the reified heritage of capitalist development."[17]

Thus Simondon's theory of concretisation suggests how technical progress might be able to address contemporary social problems through incorporating the wider contexts of human and environmental needs into the structure of the machines themselves. These secondary instrumentalisations, such as demands for safer working conditions, less severe environmental impacts, etc., are not extrinsic to the logic of technology. Rather, they respond to the inner tendency of technical development to construct synergistic totalities of human, natural and technical elements—totalities, Feenberg argues, which are disrupted within capitalism.

To repeat, Feenberg's hope is that this model of technology, and the model of resistance it suggests, can be generalised to analyse and de-reify a range of capitalist institutions. Such generalisability is possible, Feenberg argues, because institutional structures throughout capitalism share a form of rationality at their heart.[18] It is to Feenberg's conception of rationality that I now turn.

Social Rationality

In day-to-day usage, the term rationality is most usually applied to action; certain actions are understood to be rational if they are undertaken in accordance with some (good) reasons, goals or the (best) available information.[19] Feenberg accepts that modern societies are not really rational in this everyday sense. However, he does argue that "something about the structure of modernity resembles the rationality of the scientific disciplines."[20] Elsewhere he suggests, "procedures that bear a certain resemblance to those of science and mathematics operate in modern societies with tremendous effects on the whole social system."[21] In short, there is something about the resemblance of certain practices and methods to those typically employed in science that seems to ground the idea of them being rational.

In keeping with earlier critical theorists, Feenberg looks to both Weber and Marx to formulate his argument.[22] In order to distinguish his own particular formulation from those of these earlier contributors, Feenberg uses the term social rationality.[23] The workings of markets and bureaucratic organisations, as well as the adoption and use of new technology, are all examples of social rationality as Feenberg understands it. With respect to markets, this rationality can be seen in the way in which exchange takes place in terms of equivalents.[24] The ascription of prices to commodities involves a quantifiable label given to them, which enables them to be exchanged at rates that make the commodities commensurable or equivalent. Feenberg also draws upon the work of Habermas to suggest that such a focus upon equivalence invokes the logic of equations and calculation, which in turn invokes ideas of legitimacy.[25] Here, Habermas "explains the astonishing coincidence of mathematical equivalence and moral reciprocity in market relations. It is this equivalence that legitimates the market and makes it seem both natural and good."[26] Feenberg contrasts such exchange of equivalents with premodern relations that consisted in the giving of gifts and more haphazard bartering.

Modern bureaucratic organisations can be seen as socially rational in the way that they create, classify and apply rules. Science is understood to operate via the basic classification and discovery of rules that operate between different elements, such as with respect to the periodic table or the classification of species. A crucial moment in Feenberg's story here is the way that modern organisations such as businesses and bureaucracies reduce events and operations to typical "cases" that can then be procedurally decided over in terms of precedent. The influence of Weber here is clear, especially in highlighting the need to quantify and regulate much of social life in order to facilitate centralised decision-making and control. Premodern societies, Feenberg argues, decided such things in terms of rules handed down, which tended to make institutional decision-making very inflexible.[27]

The regulation and quantification of social life also facilitates another element of Feenberg's account, namely optimisation or maximisation, such as the optimisation of effort or maximisation of profit. The link to science is again made by considering mathematics and the possibilities for both measuring and maximising different ways of doing things. Similarities are once more drawn to modern business and bureaucracies. Feenberg concedes that this is something that has always happened to some degree, but argues that what makes modernity different is the scale and the centrality of such concerns to modern organisations. Much of this is only possible, of course, because of technology making such calculation and measurement possible.

Such socially rational features of markets, organisations, and technology are important, Feenberg argues, not only because of the unprecedented scale with which each features in our everyday lives, but also, crucially, because each injects a normative dimension. Markets, in which equivalents are exchanged, are perceived as fair or legitimate; once calculated or measured, a variety of different issues appear as beyond dispute; once under control by impartial experts drawing upon precise laws and rules there is no role for the input of others of for contestation. Feenberg's main point in relation to each principle is that, given this appearance of (social) rationality, critique would seem to be confined to romantic or fundamentalist reversion (to premodern utopia, to the romantic opposition of life vs machine, etc.), or what he terms social critique.

For Feenberg, then, social rationality essentially refers to three principles: exchange of equivalents; classification and application of rules; the optimisation of effort and calculation.[28] With respect to each, he suggests, this appearance of rationality depends upon their resemblance to the methods and practices of science. However, exactly why such features should seem rational simply because they resemble science is not really discussed by Feenberg. Moreover, there is little discussion of either what science involves that might make it the benchmark for rationality, or what the connection might be between his three principles.[29] In the absence of such discussion the attribution of rationality by way of these principles seems on the face of it rather ad hoc, despite the fact that these principles do seem to carry intuitive appeal. I want to argue, however, that Feenberg's intuitions can be grounded by considering a particular conception of science. In particular, I want to suggest that realist conceptions of science, following from the work of philosophers of science such as Harré and Bhaskar, suggest a better motivation for Feenberg's principles whilst being compatible with his, and critical theory's, main concerns.[30] It is to this that I now turn.

Science, Ontology and Isolatability

Feenberg is keen to point out that his main interest is technology, not science. However, his comments about rationality suggest that he believes science to be broadly successful under its own terms and that this success appears to bestow some form of apparent legitimacy not only on science itself but also on other institutions that resemble science or its practices.[31] Much, however, hinges on the conception of science implied. The conception I defend here is one that I believe critical theorists in general, and Feenberg in particular, might be sympathetic to, especially given its development in opposition to any form of positivism.[32] In very general terms, I understand science to be any activity aimed at attempting to systematically formulate knowledge of the world. Of course, at this level of generality many different academic disciplines, and indeed much of our daily activity might be understood to be some form of science. To say more, requires us to speak of particular sciences, which in turn requires us to say something about the nature of the reality under investigation by a particular science, and the conditions under which particular methods may be more or less successful.

At a very general level of analysis, science is concerned with a world of structured *things* with powers and mechanisms. Moreover, these powers and mechanisms operate in a way that is typically out of phase with the events and states of affairs through which science comes to know about them.[33] By out of phase I mean that, typically, experiences are not reducible to events and events are not reducible to the mechanisms and powers that generate them, so that knowledge, such that it is, is always an achievement. Scientific accounts become more than simple empirical descriptions when they refer to properties (powers and mechanisms) that endure *across* different empirical events or states of affairs, that is, they exist and operate *transfactually*. Thus, for example, gravity endures even though the falling leaf rarely falls straight to the ground with a constant rate of acceleration.

On this account, scientific knowledge is not arrived at primarily through deduction or induction, but through retroduction. By this, I mean that the essential moment in science is a movement from one kind of thing to another, from events (states of affairs, regularities, surprises, etc.) to mechanisms (structures of things, networks of interconnections, systems, powers) that, were they to exist, would be capable of generating the phenomena under investigation. A full explanatory account will typically involve a range of different cognitive materials, including analogy and metaphor, to construct a theory of a mechanism which, if it were to work as postulated, would account for the phenomenon in question. The mechanism suggested will then be subject to all kinds of empirical scrutiny in the context of competing explanations.

If we accept something like this conception of science and the nature of reality it is intended to illuminate, how or under what conditions is it the case that science could be perceived to be beyond challenge? I believe that science is actually never beyond challenge. Science must always be fallible, and to suggest otherwise would be to cede science to positivism.[34] There is a domain of reality, however, in which the operation of science may encourage the view of it as often, and highly, successful. This is the well-controlled experiment. Essential here is the ability to isolate real world mechanisms and correlate their triggering event with their effects. So doing makes it possible to make law-like statements of the form "if X then Y," or regularity statements (about constant rates of fall, the temperature at which boiling commences, etc.). But such interventions succeed in grounding law-like statements only if two conditions hold. First, it must be possible to activate structures or mechanisms that operate in much the same ways under the same conditions. In other words, the components of the mechanism are essentially *atomistic*. Second, it must be possible to *insulate* such mechanisms from countervailing mechanisms. If both conditions hold we have, in Bhaskar's language, a closed system.[35]

To explain the fact that knowledge gained under experimental conditions appears to be usefully applied outside of such conditions (where precise regularities of events does not occur) seems to require that knowledge of powers and mechanisms takes the form of tendency statements. On this account, the mistake of positivism is to generalise the domain of successful controlled experiments and to view science as the pursuit of laws and knowledge consisting primarily of correlations or patterns in events. Instead, the realist conception suggested by Bhaskar and others is one in which science retroduces to powers and mechanisms, which give knowledge, at best, in terms of tendencies and dispositions. That experiments sometimes work and are often replicatable suggests that some mechanisms, which have causal impacts on our world, are at least temporarily isolatable. And this is required for any science to formulate law-like statements.

Of course there is much more to the story than this, but for current purposes, the main point to make is that the kind of status that Feenberg and others attribute, even implicitly, to science depends upon the success of experiments which in turn rely on the real existence of relatively closed or closable systems, and so upon some kind of isolatability of the mechanisms and powers of interest. In other words, this status is not directly a property of science per se, but of science's operation under very particular circumstances.[36]

The positivist mistake of believing reality to be given in experience and so effectively flattening out the different layers of reality leads to an insistence that there is only really one method (deductive logic and mathematics) that is applicable to all situations and for all truly scientific disciplines. A good illustration of this mistake can be made by considering other disciplines that have sought the credibility of natural science, in positivistic terms, without such isolatability being available. The example I am most familiar with is economics, especially mainstream economics. In aping natural science, it has adopted the explanatory structure of deductivism and employed mathematics in the "explanation" of just about everything. However, such a strategy presupposes a level of isolatability in social reality that simply does not exist; the phenomena of interest to economists are not typically isolatable at all. Thus economics, as a discipline, continues to commit itself to nonsensical assumptions simply to get the modelling project off the ground, and yet it is still unable to generate the kind of law-like statements found in the natural sciences.[37]

Now, I believe that Feenberg encourages a form of this positivistic mistake when he argues that certain aspects of the social world appear rational because of the way they resemble science. In particular, those aspects of science picked out by Feenberg are those of mathematisation and deduction that are typical of positivist descriptions of "proper" science. To the extent that Feenberg does make this mistake, however, it seems easily rectified. Each of Feenberg's principles involve some kind of "if X then Y" formulation, and as such can be understood to be features that operate in roughly closed systems. In other words, instead of talking of these principles as rational because they resemble science, Feenberg could argue that his principles resemble the kind of science that can be pursued when *isolatability* is present, and so closed systems are a possibility.

Technology, Rationality and Isolation

If this much is correct, it remains to ask what implications follow from adopting this different grounding of Feenberg's principles of social rationality. I primarily focus upon Feenberg's account of technology here at least in part because it is presented as the basic model that can be extended to other institutions of capitalism.

Like Feenberg, I believe that science and technology are not the same things. Science is not simply some pure version of technology, nor is technology simply applied science.[38] Rather, technology is at least in part constituted by the technical objects that are the material results of study, including scientific study, that form the lived, experienced world. However, I want to argue, closed systems and isolatability are essential for both technology and for (experimental) science. As noted, for science, isolatability underlies the success of controlled experiments, and the use of deductive language, especially mathematics, to formulate the knowledge gained in terms of law-like statements. But what role does isolatability play in an account of technology?[39]

Also like Feenberg, I have defended a conception of technology that distinguishes different moments of the process by which technological objects come into being, one in which aspects of the world are isolated and refined and one in which they are recombined and positioned to form working devices or systems.[40] However, for Feenberg, the emphasis in making such a differentiation is upon the idea that such moments are only analytically separable. Indeed, in more recent accounts, Feenberg has spent much time warning of the danger of making too much of this separation. For him they are analytical distinctions and the problem of capitalism is that it works in such a way as to accentuate the distinctions, or make them too rigid.[41] As noted, Feenberg's solution is to democratise technology, which effectively involves reuniting the moments as much as possible, so that the concerns, values, and aesthetics of the users are as influential as possible upon the choices made by designers of technology, which crucially are always underdetermined in terms of efficiency.

My emphasis, however, is more ontological. My equivalents to his primary and secondary moments can be distinguished on the basis of different modes of existence of the phenomena that are central to them. Much of secondary instrumentalisation—concerning the ways in which artefacts and devices are positioned (enrolled) in our day-to-day lives—is concerned with phenomena that are inherently processual and internally related, and so not open to the kind of isolation feasible at the primary stage.[42] In contrast, isolation is not only feasible at the primary stage, but crucial. It is crucial, much as for experimental science, in providing the conditions for the construction of law-like statements about a series of powers and mechanisms that allow technologists to formulate understandings of different functional properties. However, it is also crucial to the way in which such properties and the devices that embody them can be recombined, providing exponentially increasing opportunities for further recombination of resulting devices, a process that underlies the familiar dynamic of technology as accelerating and pervasive.[43]

This isolatability at the primary stage underlies the applicability of mathematics, deductive logic, and mechanical recombination of isolated powers and mechanisms. As such, this moment, because of the isolatability of powers and mechanisms upon which it depends, clearly resembles a particular kind of, often very successful, science and no doubt goes some way to explaining the perceived status of technology as scientific, rational or beyond challenge. But it is only a moment in technology's coming into being. For new devices and mechanisms to be realised, or to *work*, they must be enrolled in particular networks of interdependencies or sets of social relations. They must be positioned in order to acquire such features as their social powers, identities and functions. Feenberg is correct to point out the rather different nature of this secondary, or integrative, aspect of the process. But the main differences lie not in their resemblances to science, but are ontological. Such secondary features may also resemble science of some kind, but they do not resemble the science concerned with isolatability and closure that seems to underlie Feenberg's appeal to rationality.

One implication of this is that it is not clear how isolation in the primary stage can be seen as per se damaging or problematic. There is much written, especially in the Heideggerian tradition, and repeated in places by Feenberg, that this separation is essentially wrong or "inauthentic." However, on the account I am defending here, the crucial feature of this primary stage is not that it is a violation (that requires some kind of re-uniting). Rather, it is a real possibility that only exists where powers or mechanisms can operate in relative isolation, allowing description of them in law-like statements and typically facilitating their combination or transformation to create some artefact or device. Isolatability is ontologically significant in that it is part of the conditions of possibility of (it facilitates) particular kinds of science and a particular moment in technology. In science this may not require any meaningful reintegration at all of the features isolated. In technology, however, Feenberg is quite correct to suggest that the realisation of some particular technological device requires meaningful reintegration and positioning. And such positioning may involve some form of compensation as Feenberg suggests, but it may not. In any case, it seems difficult to argue that the main issue is one of authenticity. More likely, the main issues with reintegration will be the construction of conditions under which meaningful positioning or enrolling can take place. At best, this will involve the investment of new technological objects, what Feenberg calls mediation,[44] in ways that are in keeping and draw upon lessons learned about how particular communities best want to live their lives. But often, pragmatic concerns are likely to dominate, such as there being enough time for different communities to position or enroll new artefacts in knowledgeable or meaningful ways before some new device or artefact comes along.[45]

Related to these ideas is one further implication. I would argue that viewing the social world as if it is isolatable, and as if closures are a pervasive feature of it, is a mistake. Thus the use of mathematics and deductive logic to capture aspects of the social world is likely to be problematic. It is for this reason that economists who attempt to model social reality as if it is atomistic and isolatable rarely succeed in any interesting ways.[46] The point to make here is that mistakenly viewing the world in this way causes problems for those viewing it this way.

Bearing this is mind, there seem to be some important limits to the way in which Feenberg's conception of technology can be generalised to such institutional forms as markets and bureaucracies.[47] The main way in which Feenberg generalises his account of technology is to notice that in markets and in bureaucracies there appear to be similar moments of isolation. In markets, the main force here is that of commodification.

Even labour, in becoming commodified, passes through a moment of isolation: "people are processed too. They are removed from the traditional domestic work context and relocated in factories."[48] Feenberg is careful to note that the similarities only go so far: "people … cannot of course be stripped of their non-productive aspects, like trees or minerals, but they can be obliged by the rules of the workplace to expose only their productive qualities at work."[49] However, such differences are likely to be very important in practice. Rather than, as with technology, a movement through moments of isolation then reintegration, in the commodification of labour, and presumably also in the reduction of individuals to "cases" in bureaucracies as well, there is no real *moment* of isolation at all, rather there would seem to be a more or less continual attempt to control human practices by enforcing the logic of isolatability and closure on human systems, even though they are inherently unisolatable and unclosable.

To be fair to Feenberg, his attempts to generalise his instrumentalisation theory to markets and bureaucracies are always accompanied by comments about "suitable qualifications" even though the details of such qualifications tends to be left rather undeveloped. Perhaps this framework of isolatability and open and closed systems may go some way to piecing in the details of such "qualifications." The main difference, at least on the account I am suggesting, is that with respect to technology, mechanisms are isolated in order to facilitate the often rather mechanistic recombination of different mechanisms to create different artefacts. The processes at work in markets and bureaucracies seem different in that there are almost continuous attempts by some to generate situations similar to closed systems, setting in play forces to atomise human individuals and to remove countervailing mechanisms for the purposes of generating power and control.

Concluding Remarks

The main focus of this paper has been upon Feenberg's conception of rationality. I have argued that, whatever the merits of Feenberg's general analysis, portraying certain features of modern societies as rational because of their resemblance to science seems a relatively weak link, and could even be seen as making the positivist mistake of equating science with the use of mathematisation and deductive logic. Instead, drawing upon realist theories of science, I have attempted to accommodate some of Feenberg's concerns in terms of ideas of isolation and closed systems. Each of the principles that lie at the heart of Feenberg's social rationality seem to be expressions of deductive logic, of the form "if X then Y," which is appropriate where closed systems hold. The reliability of some science and the rationality it seems to indicate emanate from situations where closures are possible, thus making it possible to formulate the kind of law-like statements that Feenberg's three principles refer to or resemble. Although grounding appeals to rationality in this way seems to have various advantages, it would also seem to suggest certain limits to the way Feenberg's position has developed. For example, generalisation of the technology framework to markets and bureaucracies may require far more qualifications than Feenberg currently employs.

However, it is worth noting that my attempts to provide an alternative grounding of Feenberg's conception of social rationality do not, I believe, undermine Feenberg's main findings or implications. For example, Feenberg's main message, that technology requires users to have a greater involvement in all stages of the design process (a call for more democratic technology), is in no way undermined by an isolation/closure perspective. It seems important to point this out, as it may be tempting to view my alternative account of Feenberg's moments as committing a serious mistake. I am not suggesting that there is simply one process in which technology is made and another in which it is used. If this were the case, Feenberg's (and my own) theory of technology would be rendered pointless, and at best would amount to suggesting the rather trivial point that technology (still left to technologists to design) should be better *used*. This, however, is not my claim. Rather, drawing upon the essential underdetermination of technology (also emphasised by Feenberg) it is clear that there are many trajectories for different technological combinations.[50] User involvement in all kinds of technological choices in as early a stage as possible seems the best way for technology to develop, in Simondonian fashion, in socially meaningful and useful ways.

In a similar vein, I would not want to drop Feenberg's resistance to neutrality conceptions of technology.[51] But, again, my emphasis is upon ontological features. In particular, I believe the main worry is the extent to which users of technology all come to see the internally related and processual social world in similar terms to the isolatable and closable world of both natural science and the primary moment of technological activity. There is little room to develop this argument here,[52] but the success of the primary moment (in generating a particular orientation to the world) under certain conditions can easily carry over (inappropriately) to the second stage of instrumentalisation, and more generally, to a variety of interactions with the social world. On the face of it, such ideas appear to come close to Heidegger's concerns about a growing tendency to see the world as standing reserves, ready to be controlled or optimised. But the emphasis, in my account, is upon the mistaken transference of ideas appropriate to the isolative moment to the rest of the world, and so the view of social reality in terms of closed systems, as atomistic and easily isolatable components, and so as resources that are ultimately predictable and controllable. On the account I defend, transferring ideas of closed systems and atomic components in this way is a mistake. It is not, as Heidegger seemed to suggest, an inevitability that we must simply find some "free relation" to. Rather, it is a mistake that is not only bad for the world, but also causes problems for those adopting this orientation to reality.

Notes

1. G. Lukács, *History and Class Consciousness; Studies in Marxist Dialectics* (Cambridge, Mass.: MIT Press, 1971).
2. Ibid., 83.
3. Max Weber, *The Protestant Work Ethic and the Spirit of Capitalism*, trans. T. Parsons (New York: Scribners, 1958).
4. Anthony Giddens, *Politics and Sociology in the Thought of Max Weber* (London: Macmillan, 1972), 44–45.
5. Although of course, whereas Weber links such unwanted aspects of modernity to Protestantism, and the quantification of nature that contributes to the secularisation of modern societies, Lukács sees the problem to lie in the needs of capitalism as roughly outlined by Marx.
6. T. Adorno and M. Horkheimer, "Towards a New Manifesto?," *New Left Review*, no. 65 (2010); Max Horkheimer, "Reason against Itself—Some Remarks on Enlightenment," *Theory Culture & Society* 10, no. 2 (1993); Max Horkheimer and Theodor W. Adorno, *Dialectic of Enlightenment* (New York: Herder and Herder, 1972). Of course the meaning of the term rationality is much debated within the accounts of critical theory. And although these debates, and indeed the precise meaning of rationality in these accounts, is beyond the scope of this chapter, I would argue that one of the most central themes in these accounts of rationality is the role of some kind of "scientific mentality" which is described by way of articulating a view of society as an object of technical control.

7. H. Marcuse, *One-Dimensional Man* (Boston: Beacon Press, 1964).
8. Herbert Marcuse, "Nature and Revolution," in *In Counterrevolution and Revolt* (Boston,: Beacon Press, 1972).
9. *Transforming Technology: A Critical Theory Revisited* (New York: Oxford University Press, 2002); "From Critical Theory of Technology to the Rational Critique of Rationality," *Social Epistemology* 22, no. 1 (2008); *Between Reason and Experience: Essays in Technology and Modernity* (MIT Press 2010); *The Philosophy of Praxis: Marx, Lukács and the Frankfurt School*, New edition. ed. (London: Verso, 2014), text; ("Marxism and the Critique of Social Rationality: From Surplus Value to the Politics of Technology.") *Cambridge Journal of Economics* 34, no. 1 (2010). Another very prominent alternative recent account of reification is provided by Honeth. However, Honeth's work does not really develop these themes. Rather it uses them as a jumping off point to something entirely different.
10. Feenberg, *Between Reason and Experience: Essays in Technology and Modernity*, 26–28.
11. Martin Heidegger, *The Question Concerning Technology, and Other Essays*, 1st ed., Harper Colophon Books Cp. 419 (New York: Harper & Row, 1977); *Being and Time* ([S.l.]: Blackwell, (1978), 1962).
12. Feenberg, *Transforming Technology: A Critical Theory Revisited*, 131–90.
13. Ibid., 192.
14. The first moment is termed that of decontextualisation and systematisation. It is because basic technical elements can be decontextualised, separated from particular contexts, that they can be combined in devices and other contexts. Thus the wood of the tree is decontextualised from the forest. Decontextualised elements must always be recombined with each other to be useful. A resulting device must be related to other devices and to some kind of working environment. Thus systematisation is the process in which these connections are established. Although this process has the potential to redress the mutilating effects of decontextualisation, such potential is severely limited under capitalism by concerns of control and profit. The second moment is termed reductionism and mediation. Technical means are abstracted by reducing complex totalities to those of their elements through which they are exposed to control. Some form of compensation is always present. Traditional craft workers simultaneously apply ethical or religious symbolism, meaning, etc. in the course of their work. Production and ornamentation proceeds simultaneously. Such mediation-centred design disappears in modern society. Here commodification dominates, aesthetics and ethical regulation are at most a constraining add-on, taking the form of packaging, indifferent to the unintended consequences of

different technologies for both humans and the natural environment. The third moment is autonomisation and vocation. The former is intended to capture the idea that the technical subject is relatively unaffected by the object on which it acts, by distancing or deferring feedback from the object. A gun is fired with the result being a small amount of kick-back, nothing compared to the effects of the firing bullet. The integrating compensation for this, vocation, captures the sense in which involvement with technical objects does act back directly upon the agent, reforming and constructing the identity of that agent. The obvious examples are premodern craft activities which served to provide a vocation, not simply from interacting with the object but from membership of a community (masons, guilds?). Such impulses are destroyed by de-skilling. The last moment, Feenberg terms that of positioning and initiative. Positioning captures that idea that we cannot transform everything to our desires. Gravity, the boiling point of water, etc. are phenomena that we cannot control but must navigate ourselves in relation to. The compensation here comes through initiative, which although curtailed in capitalism, need not be. Co-operative, non-alienating work projects, would seem to suggest a beneficial role for such initiative. A crucial part of any democratising move.

15. Feenberg, *Transforming Technology: A Critical Theory Revisited*, 198.
16. Gilbert Simondon, *On the Mode of Existence of Technical Objects*, trans. Cecile Malaspina (Minneapolis, MN: Univocal Publishing, 2010).
17. Feenberg, *Transforming Technology: A Critical Theory Revisited*, 187.
18. Actually, Feenberg's arguments for generalisation take two forms. In some places he argues that technology is now an integral part of all human systems. Whereas for Marx the only really fully-blown example of technological development lay within the factory, so it was excusable for Marx to limit his analysis about reificatory systems and emancipatory change to the factory floor. However, today, given that technology has spread to almost all aspects of human societies the potential for emancipatory de-reification requires a far broader focus.
19. Of course the term has been given a variety of different, more specialist meanings within different disciplines. For example, within economics it is often given a very formal meaning in terms of various axioms of logical consistency.
20. Feenberg, *Between Reason and Experience: Essays in Technology and Modernity*, 156.
21. "Marxism and the Critique of Social Rationality: From Surplus Value to the Politics of Technology," 40.
22. From Weber, Feenberg takes the idea of rationalisation and its focus upon the increased roles for calculation and control in modernity. Moreover, Feenberg highlights the important feature of rationalisation is its relation

to science. For example, he retains, from Weber, "the emphasis on forms of thought and action that bear some resemblance to scientific principles and practices" *(Between Reason and Experience: Essays in Technology and Modernity*, 158). Feenberg's debt to Marx is less clear. Although positive references to Marx abound in Feenberg's work, he clearly rejects Marx's value theory, and possibly his theory of commodity fetishism. The one thing that Feenberg explicitly adopts from Marx is what he terms a rational critique of rationality, which is more an approach to critique rather than the substance of Marx's contributions ("Marxism and the Critique of Social Rationality: From Surplus Value to the Politics of Technology.") In essence, this critique involves uncovering the biases that remain in socially rational systems. Although crucial to Feenberg's contributions, I shall have to ignore this here. What interests me, rather, is the form of rationality that underlies Feenberg's assessment of capitalism.

23. See especially "Marxism and the Critique of Social Rationality: From Surplus Value to the Politics of Technology."
24. This idea does indeed seem to be central to Marx's conception of exchange, especially in his attempt to locate the source of profit. It is only if markets create situations under which all things exchange as equivalents that it becomes a mystery to explain how profits are generated.
25. J. Habermas, *Towards a Rational Society* (Boston: Beacon Press, 1970).
26. Feenberg, "Marxism and the Critique of Social Rationality: From Surplus Value to the Politics of Technology."
27. *Between Reason and Experience: Essays in Technology and Modernity*, 159.
28. Ibid., 159; "Marxism and the Critique of Social Rationality: From Surplus Value to the Politics of Technology."
29. Although Feenberg has given various accounts of science, he is typically more concerned with the differences and similarities between technology and science, rather than those features of science that might actually ground his conception of social rationality; for example see "Science, Technology and Democracy: Distinctions and Connections," http://www.sfu.ca/~andrewf/Science%20and%20Democracy2.htm (2009).
30. R. Bhaskar, *A Realist Theory of Science* (Brighton: Harvester, 1978); R. Harré and E. Madden, *Causal Powers* (Oxford: Basil Blackwell, 1975).
31. Feenberg, *Between Reason and Experience: Essays in Technology and Modernity*, 157.
32. Which of course critical theorists have been particularly critical of; see for example Max Horkheimer, "The Latest Attack on Metaphysics," in *Critical Theory; Selected Essays* (New York: Herder and Herder, 1972). Just to be clear, my understanding of positivism is as, in the first instance,

a theory of knowledge, one in which knowledge is given in experience. As knowledge must be of events, so any general knowledge that might be possible can only be of correlations of events (typically understood as laws). Thus because events are given in experience, reality is given in knowledge, so that knowledge is certain and grows in a linear way, by accumulation.

33. There is little space to defend such an assertion here, but see Bhaskar *A Realist Theory of Science.*
34. See note 36.
35. See in particular Bhaskar *A Realist Theory of Science.*
36. Of course much has been said about the social irreducibility of science as a practice—experiments are work, interventions by people making choices about what to study, constructing conceptions and interpretations as they go, etc. However, none of this undermines (and indeed it is made intelligible by) the fact that without an ontology of isolatability, it would not be possible to explain why science takes such different forms in different circumstances.
37. Moreover, the insistence on attempting to formulate knowledge in this way makes it impossible to obtain the kinds of knowledge of the social world that is available. Of course, it is easy to see why such knowledge might be wished for. Finding some regularity, which can easily be transposed into the 'if X then Y' form, allows fairly simple recommendations of policy and control of the economy or some other aspect of the social world. However, knowledge of the social world has never been obtained which mirrors that of the non-social world, and given everything we know about the social world (for example, its internally related and processual nature) there seems no reason for supposing that the required closed systems are possible to construct. Although some may challenge this assessment of the state of affairs in modern economics, few would suggest that the status of experiment in economics closely resembles that in the natural sciences.
38. Although, of course, much depends upon what is meant by science and by technology. For a detailed account of these issues, see Clive Lawson, *Technology and Isolation* (Cambridge University Press, 2017). see especially Chaps. 2–4.
39. Although there is little space to develop an answer to this question here, this is the main focus of the Lawson book, Ibid.
40. "Technology, Technological Determinism and the Transformational Model of Technical Activity," in *Contributions to Social Ontology*, ed. Clive Lawson, John Latsis, and Nuno Martins (London: Routledge, 2007); "An Ontology of Technology: Artefacts, Relations and Functions," *Techné: Research in Philosophy and Technology* 12, no. 1

(2008); "Ayres, Technology and Technical Objects," *Journal of Economic Issues* 43, no. 3 (2009); "Technology and the Extension of Human Capabilities," *Journal for the Theory of Social Behaviour* 40, no. 2 (2010); *Technology and Isolation.*

41. Feenberg, *Between Reason and Experience: Essays in Technology and Modernity*, 74.
42. For my account of social ontology in this context, see Lawson, *Technology and Isolation.* Chap. 2.
43. Again, there is little space to develop such arguments here, but for a detailed account see ibid., Chap. 9.
44. Feenberg, *Between Reason and Experience: Essays in Technology and Modernity*, 111.
45. Lawson, *Technology and Isolation.* Chap. 9.
46. Additionally, although there is little space to develop the argument here, I have argued elsewhere that mistakenly attempting to interpret the social world as isolatable and closable explains why those on the autism spectrum find social interaction so challenging. See ibid., Chap. 9.
47. Feenberg, *Between Reason and Experience: Essays in Technology and Modernity*, 168.
48. Ibid., 170.
49. Ibid.
50. Essentially, the idea of underdetermination is that at any moment there is more than one technically feasible solution to any problem, and that the choices about which solutions to pursue and develop, remain social. See for example, Feenberg, ibid., 10.
51. Where I take neutrality views to involve the idea that there is nothing in general that can be said about the nature of technology, and that any problems or advantages of particular technologies only come with the way they are employed in some particular circumstance.
52. Although see Lawson, *Technology and Isolation.*, Chap. 11.

Bibliography

Adorno, Theodor and Max Horkheimer. "Towards a New Manifesto?" *New Left Review*, no. 65 (Sep–Oct 2010): 33–61.

Bhaskar, Roy. *A Realist Theory of Science.* Brighton: Harvester, 1978.

Feenberg, Andrew. *Between Reason and Experience: Essays in Technology and Modernity.* MIT Press 2010.

———. "From Critical Theory of Technology to the Rational Critique of Rationality." *Social Epistemology* 22, no. 1 (2008): 5–28.

———. "Marxism and the Critique of Social Rationality: From Surplus Value to the Politics of Technology." *Cambridge Journal of Economics* 34, no. 1 (Jan 2010): 37–49.

———. *The Philosophy of Praxis : Marx, Lukács and the Frankfurt School*. New edition. ed. London: Verso, 2014.
———. "Science, Technology and Democracy: Distinctions and Connections." http://www.sfu.ca/~andrewf/Science%20and%20Democracy2.htm (2009).
———. *Transforming Technology: A Critical Theory Revisited*. New York: Oxford University Press, 2002.
Giddens, Anthony. *Politics and Sociology in the Thought of Max Weber*. London: Macmillan, 1972.
Habermas, Jürgen. *Towards a Ratioanl Society*. Translated by Jeremy J. Shapirio. Boston: Beacon Press, 1970.
Harré, Rom and Edward H. Madden. *Causal Powers*. Oxford: Basil Blackwell, 1975.
Heidegger, Martin. *Being and Time*. [S.l.] Translated by John Macquarrie and Edward Robinson. Oxford: Blackwell, (1978), 1962.
———. *The Question Concerning Technology, and Other Essays*. 1st ed.Translated with an Introduction by William Lovitt. Harper Colophon Books New York: Harper & Row, 1977.
Horkheimer, Max. "The Latest Attack on Metaphysics." In *Critical Theory; Selected Essays*, xxi, 290 p. New York: Herder and Herder, 1972.
———. "Reason against Itself—Some Remarks on Enlightenment." *Theory Culture & Society* 10, no. 2 (May 1993): 79–88.
Horkheimer, Max, and Theodor W. Adorno. *Dialectic of Enlightenment*. New York: Herder and Herder, 1972.
Lawson, Clive. "Ayres, Technology and Technical Objects." *Journal of Economic Issues* 43, no. 3 (Sep 2009): 641–59.
———. "An Ontology of Technology: Artefacts, Relations and Functions." *Techné: Research in Philosophy and Technology* 12, no. 1 (2008).
———. *Technology and Isolation*. Cambridge: Cambridge University Press, 2017.
———. "Technology and the Extension of Human Capabilities." *Journal for the Theory of Social Behaviour* 40, no. 2 (2010): 207–23.
———. "Technology, Technological Determinism and the Transformational Model of Technical Activity." In *Contributions to Social Ontology*, edited by Clive Lawson, John Latsis and Nuno Martins, 32–49. London: Routledge, 2007.
Lukács, Georg. *History and Class Consciousness; Studies in Marxist Dialectics*. Translated by Rodney Livingstone. Cambridge, Mass.: MIT Press, 1971.
Marcuse, Herbert. *One-Dimensional Man*. Boston: Beacon Press, 1964.
———. "Nature and Revolution." In *In Counterrevolution and Revolt*, p. 138. Boston: Beacon Press, 1972.
Simondon, Gilbert. *On the Mode of Existence of Technical Objects*. Translated by Cecile Malaspina. Minneapolis, MN: Univocal Publishing, 2010.
Weber, Max. *The Protestant Work Ethic and the Spirit of Capitalism*. Translated by Talcott Parsons. New York: Scribners, 1958.

PART II

Between Democracy and a Politics of Resistance

CHAPTER 6

Transforming Dystopia with Democracy: The Technical Code and the Critical Theory of Technology

Graeme Kirkpatrick

Introduction

This chapter is centrally concerned with Andrew Feenberg's[1] notion of a "technical code." According to Feenberg, all societies contain "technical elements": individual bits and pieces of technology that are brought together in specific concatenations to form artifacts that people recognize and use.[2] The way that this is done in each society is consistent across technical practices in that it corresponds to the prevailing notion of what technology is and how we are supposed to go about "doing" it. Performing technology in this way is a scripted activity in which people and objects come together to produce both individual technology designs and, at the same time, to reproduce the prevailing idea of what technology is. The technical code is the script.

The technical code that constrains, guides, and is used by those involved in design is not a neutral set of rules but reflects the dominance of particular social interests, while the resulting designs re-enforce

G. Kirkpatrick (✉)
University of Manchester, Manchester, UK

D.P. Arnold and A. Michel (eds.), *Critical Theory and the Thought of Andrew Feenberg*, DOI 10.1007/978-3-319-57897-2_6

their dominance. Much of the terminology of political discourse tends to be shared among the parties to a debate, but some articulations of them appear more coherent and truthful than others, so some designs will be favored within the framework set by the dominant articulations of the technical code. These designs are selected by what Feenberg calls "hegemonic technological rationality," which is manifest in articulations of the code specific to modern capitalist societies.

The technical code is always subject, however, to the possibility of challenge and contestation. Like democratic politics itself, it rests on contingency and is open to a range of competing participants who struggle for control of the code. In any given combination of technical elements, a set of values are condensed with strictly technical requirements and the outcome reflects dominant social interests. Whereas constructivism sees only the relatively innocent practice of labeling by "socially relevant groups," which "shapes" technologies by inscribing them with meanings that comport with the practices of specific groups of users, Feenberg's lens is wider and his political focus is sharper. Disputes over the shaping of any particular technology may have implications for the meaning of technology itself and thus, he argues, these contests over design ultimately hold out the possibility of a change of civilizational model.

The idea of such profound and the far-reaching transformation was for a long time unthinkable within the tradition of critical social theory from which Feenberg writes. For Herbert Marcuse[3] and other Frankfurt theorists, technology was essentially implicated in the modern condition. As the physical incarnation of a pervasive instrumentalism that menaced meaning and threatened to leave us all at the mercy of experts whose only concern was efficiency, technology was beyond reform. Technocracy was, as Feenberg points out, the dystopian cultural backdrop for that generation of critical theorists and all their efforts were devoted to defending the possibility of non-technological meaning-making as the true human vocation.

I will suggest that Feenberg's break with the technocracy dystopia, which is made possible in part by his engagement with and transcendence of constructivism, is incomplete. For this reason, he understands the central antagonism of technical politics in terms of the opposition of democratization of technology design to its subordination by hegemonic technological rationality. I will argue that the idea of hegemonic technological rationality resembles the old technocracy too closely. This means that Feenberg underestimates the scope of technology reform. As the place of technology on the cultural horizon has shifted, so what we mean by it has changed and this has implications for the range of possible counter-articulations of the technical code. At the same time, while democracy is the value that makes these articulations possible, Feenberg tends to exaggerate the extent to which technology can be democratized: what he calls the boundary of technique can be moved but is, I argue, an incorrigible feature that sets limits to mass participation.

I begin with an account of Feenberg's arguments regarding the dystopia of technocracy and his reasons for abandoning that version of the technological imaginary. This involves an account of his concept of hegemonic technological rationality and the related concepts of "the boundary of technique" and the "technical illusion," which enable us to see how his technical politics both presupposes and moves beyond constructivism. Section "The Technical Code" turns to the central concept at issue, which is the idea of the technical code. Here the focus is on Feenberg's account of how values get impressed on technology at the scene of design, where the technical code "condenses" values and technical considerations in specific designs. This has implications for technology reform, which cannot proceed in an instrumental fashion[4] and equally cannot be a matter of (utopian) wishing nicer machines into existence. I will try to locate the technical code more firmly in politics, on the side of contingency and articulation,[5] in order to clarify the sense in which changes to the technical code include and perhaps presuppose an ontological, evental moment.

Section "Transforming Technology" explores Feenberg's thesis of technology transformation in light of this. Non-instrumental technical politics involves agency in relation to the technical code in three distinct forms. First, democratizing design exposes more people to the ontological reverberations of technological intervention, fostering a more circumspect approach to design, and more attention to the remediations that make technical action meaningful. Second, the openness of the technical code to heterodox articulations creates the potential for designs that subvert the hegemonic technological rationality, broadening what we mean by "efficient machines." Finally, Feenberg writes about the mediation of values from the current civilizational horizon through the technical code. As technical politics changes what we mean by technology, so the range of values available to be drawn upon expands, heralding a civilizational shift.

The final section attempts a critical appraisal of the idea of the technical code and the work this concept does in Feenberg's critical theory of technology. I suggest that democracy carries much of the argument in connection with the three dimensions of transformational practice, integrating them into his vision of the direction of civilization change. This is empirically problematic, since democracy when applied to technical systems has not always led to their ethical improvement when viewed in historical perspective. Finally, I will argue that this difficulty reflects the enduring hold of the technocracy dystopia on Feenberg's thinking. The constellation within which we appraise and understand the meaning of technology has changed in a manner consistent with his theory, but in doing so it has undermined the purchase of a technical politics that opposes democracy and difference on one side to hegemonic technological rationality on the other.

After Technocracy

Earlier critical theorists aligned technology with technocracy and the menace of instrumental reason. Feenberg argues that we have now moved beyond this and the "dystopian" technological imaginary should be abandoned. Of particular importance here is the extent to which human individuals now have agency within the technical sphere. This contrasts with the situation in other areas of culture. As Feenberg notes, "despite discouraging developments in other domains, agency in the technical sphere is on the rise."[6] He associates this with changes to the technical infrastructure itself:

> We are witnessing the end of dystopia as the defining technology of our time shifts from great centralized systems such as electric power and broadcasting to the more loosely structured world of the computer.[7]

In place of technocracy, which was monolithic, Feenberg suggests contemporary power is operative in the shaping and implementation of technology through what he calls the hegemony of technological rationality.

Drawing on constructivism, Feenberg shows that technology design and use is often contested rather than top-down and controlled by experts. Constructivism shows that design decisions are underdetermined. That is, for any problem a technological capability might be used to solve there will be more than one equally feasible design. Social actors select among the available alternatives and in this way technology is shaped by, rather than determining of, social relations. For constructivists, social shaping is normally a matter of a distinctive vocabulary applied to technology in its development phase that comes to determine our perception and understanding of it. Social groups prefer some descriptions to others because those descriptions tend to produce artifacts that comport with their interest in the technology. Once a description takes hold, artifacts are re-shaped to fit, so the shaping is both symbolic and material.[8]

Feenberg points out that these contests are pre-configured by a deeper structuring, specific to technology as a social practice. This structuring overdetermines the choices made by those involved in the design process. Hegemonic technological rationality ensures that some values (and corresponding social interests) are more likely to prevail than others: "Since technology is not neutral but fundamentally biased toward a particular hegemony, all action undertaken within its framework tends to reproduce that hegemony."[9] The hegemonic technological rationality ensures that the values of managerial control, a narrowly defined idea of efficiency, and a stern absence of other values are pervasive at the scene of design.

This hegemony is what ensures that all our technology is recognizably technology: it overdetermines what we will accept as "real" technology and, in so doing, re-enforces existing forms of social power. In this way, hegemonic technological rationality sets what Feenberg calls the boundary of technique. The boundary of technique is the inscription of necessity in technologies; the sense that they must be accepted and complied with because they embody the correct technical solution. Whereas in technocracy this boundary was set and fixed, in the sense that few would ever have thought to question a machine beyond learning how it worked, in recent decades increased the willingness of diverse social agents to challenge technical designs and subvert authorized patterns of use have created instability. For this reason,

> ...the boundary of technique is never clear. In fact, identifying that boundary is one of the most important stakes in the struggle for and against alienated power.[10]

Democratic technical politics aims at shifting the boundary of technique, making more technical practices open to exploration and experiment and, in the process, changing our conception of what technology is. The extent to which agency is now present in the technical sphere means that, for Feenberg, we can now speak of "technical politics." This is a development of the constructivist thesis that different social groups may challenge technology design and that these contests "shape" the resulting artifacts. It adds to that argument the observation that when they are successful democratic interventions in technology design can change what we mean by technology, moving the boundary.

At the same time, progress in technical politics disabuses us of what Feenberg calls the "technical illusion." Technology is widely perceived as liberating because it appears to free us in various ways from tricky world-entanglements. While acknowledging the efficacy of technology—it really does enable us to solve problems and in this sense, it is liberating—Feenberg emphasizes that there is also an illusory element to this. Technology seems to inoculate us against the negative blowback caused by our manipulative dealings with the world. By opening up technical practices to a wider range of social logics than hegemonic technological rationality, democratic technical politics counters this illusion. The emphasis on technology's negative impacts in this argument reflects Feenberg's distinctive ontology of technology, which, I will argue, inhibits his development of the idea of radical technical politics.

The Technical Code

Feenberg's idea of the technical code is centrally concerned with how values get impressed on artifacts at the scene of design and the related question of how the meaning of technology as a whole comes to be aligned with the interests of dominant social groups. He agrees with constructivism that technology is shaped by social processes all the way down: there is no "inviolate level"[11] of purely technical determination that ensures any particular design will be selected.[12] Most constructivists make the issue of which design is chosen a function of the success of a "relevant social group." However, for Feenberg, this abstracts our understanding of social shaping from wider social networks and webs of meaning. Indeed, some constructivist studies even lose sight of anything that connects the disparate instances of technology as technology, such is their focus on the sociological minutiae of each case.[13]

The technical code "…is the realization of an interest in a technically coherent solution to a general type of problem"[14] and, as such, it is manifest wherever social interests are at work shaping technology designs. But the resulting designs bear the impress of the technical code, and as a result their design also re-enforces the current social hegemony. As well as informing the actions of proximal social actors, each manifestation of the code "…serves as a paradigm or exemplar for a whole domain of technical activity."[15]

The "code" metaphor conflates three sets of ideas. First, "it is most essentially the rule under which technical choices are made in view of preserving operational autonomy."[16] Here the emphasis is on the routinization of design practice so that choices result in artifacts that support and re-enforce existing power structures. Second, artifacts are shaped to a certain end, but they also communicate their function to other actors and enroll them in relevant activities.[17] Feenberg writes that "organizations must encode their technical environment, not merely associating technology with certain signifiers, but installing these signifiers in its very structure."[18] Finally, the technical code is a discourse in the Foucauldian[19] sense: "the technical code… is a 'regime of truth' which brings the construction and interpretation of technical systems into conformity with the requirements of a system of domination."[20] Here the technical code extends beyond local sites of technology design and is inscribed in norms of perception and practice that pertain to technology, but are operative over wide social domains.[21]

Feenberg writes that "Modern societies… build long networks through tightly coupling links over huge distances between very different types of thing and people."[22] The technical code runs through these networks, controlling the description, and shaping of new artifacts where it is a rule of participation; the appearance of technology qua technology, where it is a kind of signification, and extending to the socio-cultural horizon, where it finds technology's place in the categorical orderings specific to society (the technological imaginary). At each of these points, the technical code is met by countervailing forces that may affect its operation. For example, a society might become "more technological," altering the place of technology on its horizon: Feenberg argues that modern societies think of themselves in this way, distinguishing themselves from traditional ones, and this difference becomes a matter of identity for people in those societies. Similarly, our expectations and perceptions of what counts as a technology may be altered as a result of ideological or political pressure. Feenberg gives the example of environmentalism, which he says has succeeded in installing the question of sustainability close to the heart of what technology means. And, as constructivists have shown, disputes arise all the time when new artifacts are in development. Feenberg's theory shows that these conflicts stand in a wider context and that the technical code, "invisibly sediment[s] values and interests in rules and procedures, designs and artifacts that routinize the pursuit of power and advantage by a dominant hegemony."[23]

The idea of the technical code is pivotal for Feenberg's thesis that squabbles over technology are important to the contemporary operation of political power. He refers approvingly to Ernesto Laclau and Chantal Mouffe's clarification of the idea of hegemony. For them, "…the concept of hegemony supposes a theoretical field dominated by the category of *articulation*; and hence that the articulated elements can be separately identified."[24] This emphasis on articulation reflects a new level of autonomy for the political from infrastructural determinants and this move to autonomize what might once have been called the "superstructure" defines their post-Marxism. Focusing on language as the medium of politics, they argue that emancipatory struggle has moved from being a matter of rival, class-based articulations of the elements of political discourse[25] to one in which articulation of difference itself is the measure of radicalism. Their theory of politics as the struggle against hegemony holds that society itself is an effect of displaced antagonism. The latter is understood in terms of the exclusion of difference in the field of contingency (which we might think of as withholding of recognition of a group or validity of a claim of unfair treatment). Hegemony is established when antagonism like this is placed out of view by power: the idea of society (as a given, beyond the reach of contingency and articulation) emerges here as a kind of horizon of political activity.

In this understanding of political struggle against hegemony, "social logics replace ontological foundations"[26] by loosening the hold of the dominant conception by re-articulating the elements of political discourse. Interestingly, however, Laclau and Mouffe exclude technology from the field of contingency and articulation. The struggle for hegemony is waged in terms of control over syntactical elements of political discourse and while the areas of life that are open to articulation in that way expand in the capitalist era,[27] it does not reach those dimensions of the social formation that are paradigmatic, or systematized. They exclude things like the "reorganization of an ensemble of bureaucratic administrative functions according to criteria of efficiency or rationality"[28] from the politics of hegemony because they are not located in contingency.[29]

Feenberg's approving reference to Laclau and Mouffe would seem to place the technical code on the side of articulation in their model and implies a development of technical politics as a contest between rival social logics and the hegemonic technological rationality. If we extend this to technology, the struggle against hegemony encompasses disputes over technology design. Just as with the terms of political discourse, hegemony in technology design works through a determinate articulation of elements. Technology design is initially an open field of rival articulations associated with creativity and diverse social logics. This becomes antagonism when the disputes are sharpened, perhaps by a sense that different designs serve opposed social interests. The hegemonic technological rationality displaces antagonism when it installs the dominant image of technology (as authoritative, the domain of expertise). When technology has been subject to hegemonic codification this is because creativity, open experimentation, and the exchange of ideas have given way to the decisive, overdetermining impress of the hegemonic technological rationality. At this point the technical code confers a seeming ontological status on the artifact, classifying it as technology (and therefore as beyond question).

Sometimes, Feenberg implies this is how he understands the struggle against hegemonic technological rationality, but he rarely employs the vocabulary of "articulation" (which seems so apt to the scenario of reconfiguring technical elements). Instead, he suggests that the technical code is not a matter of mere social logics, but an ontological category too:

> The technical code has (social) ontological significance in a society where domination is based on control of technology. It is not merely the rule under which means are chosen to achieve certain ends. Much more than that, it is the principle of organizational identity and survival.[30]

While he takes the first steps towards conceiving technical politics as a part of the radical democratic politics of contingency, Feenberg does not go all the way. On his reading of the hegemonic technological rationality, its ontological dimension exceeds the range of possible subversion by rival social logics and the technical politics of articulation.

If we think opposition to the hegemonic technological rationality in terms of articulation, technical politics becomes a matter of introducing different social logics into the operation of the technical code.

We can see this in some of Feenberg's examples, such as his account of Minitel users who transformed the French proto-internet from a mere electronic phone book into a kind of dating agency.[31] In so doing, they shook the equipment free from its symbolic association with organization and efficiency and broadened its range of communicative uses. They also, perhaps, contributed to the development of an idea of technology as something to be played with, or used playfully, familiar to us now (and very different from the dystopic technological imaginary).

However, when he assigns it "(social) ontological significance," Feenberg reads the hegemonic technological rationality as part of the ontological difference between modern and other societies. The hegemonic technological rationality works according to "three main principles"[32]: the exchange of equivalents; classification and application of rules, and calculation of optimal results. In traditional societies, these were not operative, so that straightforward domination and informal rules of thumb were more salient.[33] The move from one to the other is paradigmatic, or concerns more profound parts of the social cultural order than can be comprehended (still less changed) politically.

Underlying this idea is Feenberg's own ontological theory, which he calls instrumentalization theory. There he states that all technology has a primary moment in which technology strips objects from their original position in the world and turns them to human purposes. The violence in this operation is compensated for in "secondary instrumentalization," which is more evident in traditional, pre-capitalist societies. Here, as if to make up for the original harm, tools are decorated and the practice of using them invested with significance. Capitalist modernity differs from earlier cultures in the way that it carries out this second moment. Rather than wrapping technical practice in meaning, modern societies emphasize its neutral character. The process of shaping technology involves minimizing its semantic content so that part of what we mean by technology is that it is merely efficient means, that is, it is not meaningful.

Feenberg's democratic technical politics is conceived as a counter to modernity's very particular codification of technology. At this point, his vision of reform extends to deep structural truths about technology, especially what he calls the "action–reaction" principle. Feenberg writes that modern societies' failure to carry through secondary instrumentalization properly results in multiple adverse consequences that are initially concealed by the illusion of technique. Technology may enhance the control of its primary users, or more often their managers, but in so doing it unleashes an inescapable reaction that plays out elsewhere, perhaps in environmental consequences on the other side of the planet, often closer to home in such things as hazardous tools deemed "cost-effective" on narrow construals of efficiency.

Technical politics, then, has two layers or aspects. The first lies within the field of articulation as just described, and involves opening up the technical codes to diverse social logics and articulations that subvert the hegemonic technological rationality. The second layer draws on an ontological conception of technology as a mode of world-disclosure (instrumentalism) and identifies neglected potential as the basis for critique. This layer of technology transformation consists in an infusion of meaning and value of technical practices.

For Feenberg, the second layer is essential to critique because it is the realization of the potential for a more meaningful world-relation that connects the otherwise disparate moments of technical politics. The disparate activities involved in conflicting articulations are unified through the idea of released potential. Once the technology has been opened up to democratic participation, critique faces the question of what it is in one articulation or "sort" that connects it to others and permits us to align them as manifestations of a single counter-strategy. The ontological layer of technical politics is Feenberg's framework for answering these questions, but I submit that it inhibits the development of his technical politics as a form of politics. Even as he correctly identifies technology's transition into the realm of the syntactical as against paradigmatic; contingency as against ontology and articulation as against unquestionable authority, his continuing attachment to the dystopian conception of modern technology as a negative relation to the world prevents him from capitalizing fully on these ideas.

Transforming Technology

Technology holds a distinct structural position in society and it has internal protocols that must be respected in any account of the reform process. This means that while the design is an important opening for political practice it must be embedded in an account of wider transformation. Perhaps part of the motivation for ontologizing hegemonic technological rationality is to constrain the democratization thesis, to keep it "realistic." While constructivism enables Feenberg to open up the scene of design, so to speak, so that we may think the possibility of people making demands and shaping technologies in a politically motivated direction, he is also wary of utopianism.[34]

For this reason, Feenberg is clear that technology design is never a process that accommodates strong sentiments, irrational ideas or inappropriate motivations.[35] If it did, this would compromise the real constraints that apply to technology, as against other spheres of creativity. What counts as rational in technology design may shift, but without violating or flouting basic conditions. Just as Adorno[36] relativizes rationality to social and historical context while maintaining that the truths of mathematics are invariant, so there is a basic kernel of attitudes and procedures that always constitutes technical reason, and hegemonic technological rationality condenses these with values in the technical code. It is not clear, however, why this technical reason should be associated with anything as unpleasant sounding as primary instrumentalization, or even touch upon the question of our world-relations.

Feenberg does hold out the possibility that, "The capitalist technical code, adjusted to the need to maximize profit and control the workforce, [c]ould be replaced by a different code that would take into account a wider range of variables."[37] Moreover, "An alternative modernity ...would recover the mediating power of ethics and aesthetics," linked to the "democratization of technically mediated institutions."[38] I will suggest that, as a result of the residual dystopianism in his ontology, Feenberg's conception of this transformation is both too austere in limiting the range of values that may feature in re-articulations of the technical code and too generous in suggesting that democratization alone will lead to desirable civilization change.

Feenberg suggests that the mere presence of more people in the design process, and the fact that technology design becomes a more transparent process as a result, will make for technology that is less harmful to nature and less prone to creating new hazards, which are often borne by weaker members of society. The first gain of democratization, then, would be to create design communities that are aware of the "threat" posed by technologies, even when they seem to enhance human "control":

> Only a democratically constituted alliance of actors, embracing all those affected, is sufficiently exposed to the consequences of its own actions to resist harmful projects and designs at the outset.[39]

When more people are exposed to and made aware of the effects of technology, even if they are codified in narrowly instrumental terms, this will create resistance internal to the design process. Exposure to and awareness of the harm latent in narrowly efficient designs might result in a degree of circumspection about technology that begins to alter the ethos of technology design. For example,

> Normally... feedback is reduced or deferred so that the subject of technical action is safe from the power unleashed by its own actions. But technology can "bite back"... with fearful consequences... Today we are most obviously aware of this from the example of climate change, an unintended consequence of almost everything we do.[40]

In some cases, we can see that defining technology in terms of narrowly construed efficiency has the effect of binding agents to negative consequences of a design. However, there is something of a leap from there to a concern with environmental blowback widely construed (as it is in these remarks).[41]

Reforming the scene of technology is a conservative move aimed at installing containment as an internal value of technological rationality. It is interesting that this is essentially a *negative* principle. Whereas Feenberg opposes conservative strategies aimed at containment of technology from positions in culture outside technology (for example, restrictions on embryo research), he acknowledges that checking certain kinds of technology design from within, so to speak, will be the primary beneficial consequence of democratization. Here Feenberg re-works an argument of his critical theory forebears, in which narrow instrumentalism is charged with neglecting both wider social consequences and human potential (outer and inner nature). The argument betrays an ongoing concern with technology as a threat, which he has inherited from dystopian thinking.

Here we can see that Feenberg overestimates what democratic reform will achieve, in the sense that he assumes more participants at the scene of design will lead to more circumspect choices undertaken with a wider sense of technology's negative impacts. At the same time, he perhaps inadvertently restricts the range of likely reforms to a largely defensive principle. The goal of radical democratic technical politics is to safeguard the scene of design so that it is free from powerful social interests. Success in this should be seen,

> ...not as the salvation of the whole, but as a protective umbrella under which social creativity can operate at the micro-level of particular institutions and workplaces.[42]

While democratic reform is intended to create an opening for new ideas and designs that embody different values, these are now restricted to options that fit the definition of enriched practices of secondary instrumentalization. The key idea here is potentiality: the social (ontological) side of the hegemonic technological rationality is matched by an idea of latent potential. Feenberg writes that "...it is important to retain a strong notion of potentiality with which to challenge existing designs,"[43] but this idea also reflects the enduring influence of earlier versions of critical theory.

Feenberg allows that resistance to dangerous or hazardous innovations just described might tip over into different articulations of the technical code in design practices. Here too, though, he is cautious. As we have seen, new potentials become available in the course of the design process that would be overlooked by the hegemonic technological rationality, but alternative technical codes formed through new articulations expressing rival social logics will "guide the design of future technology,"[44] producing different rules for the selection and concatenation of technical elements. What is protected under the umbrella and set free by different articulations of the technical code is presumed to be a kind of positive potential.

This question of potentiality is closely related to the theme of the agency at the scene of design. Designs shaped by the hegemonic technological rationality's articulation of the technical code will tend to discourage agency and to re-enforce the impression that technology is beyond our control. But as more people become involved and are able to fuse their demands with technical requirements, so the boundary of the technique is moved back, so to speak. Different articulations of the technical code make this possible and the expanding margin for agency Feenberg notes in diverse contemporary technical settings is a sign that it has already happened. What is in doubt, however, is that anything corresponding to human potential has been liberated. Rather, the observable changes merely concern reconfigurations of multiple sets of relationships and networks.

The context in which these kinds of transformation are embedded concerns the wider value-horizon of technology development, which is set by the kind of society that is being produced from day to day by social agents, including people making and using technology. Feenberg envisages technological transformation that extends to this level when he writes:[45]

> In a technical democracy, technical work would take on a different character. Design would be consciously oriented towards politically legitimated human values rather than subject to the whims of profit-making organizations and military bureaucracies. These values would be installed in the technical disciplines themselves, much as the value of healing presides over biological knowledge of the human body in medicine.[46]

Here, his vision is of technical politics embedding the technical code in a different kind of world-ordering or (social) ontology—a more profound shift than is comprehended by the notion of multiple competing social logics. This paradigm shift would expand the range of values that can make their way into technology design.

The importance of the notion of potential in Feenberg's critical theory is clear. We can conceive technical politics as a process of transformation that extends from contests within the design, where articulation is the key concept, to a more profound transition in which society moves from treating the world as raw materials (efficiency) to a different world-relation in which human potential is liberated. My point in this paper is that we can and do have changes of the first kind without any corresponding advance in terms of the second. The thesis of democratization and efforts to move the boundary of technique are, perhaps, desirable in themselves, but they need not bring about anything like a civilization changing liberation of human potential.

As we have seen, for Feenberg the meaning of technology has changed as we have left the "dystopian" constellation that preoccupied his critical theory forebears. This seems to coincide with digitization (though we cannot claim that digital technology caused the change), and the result is a changed relation between technology and values. The social logics that might articulate the technical code have become more diverse and the code's entanglements with power more complex. It follows that the fault lines, where contests over technology design occur that shape the meaning of technology, have also moved. Feenberg acknowledges this, but his model for technical politics remains premised on a confrontation between narrow, limiting, dominant technological rationality on one side, and potential-releasing, democratic participation on the other.

As we have seen, a central part of Feenberg's indictment of capitalist modernity is that it only allows one value—efficiency—to inform secondary instrumentalization. His emphasis on hegemonic technological rationality leads him to posit democracy as the privileged counter-value, through which space might be created for rival social logics to articulate the technical code. This leads to a "widening" of the range of applicable values, and ultimately this makes civilizational transformation feasible. However, the move to a constellation in which other values might enter technical practices need not in itself comport with any vision of "progressive rationalization" or indeed deepening democracy; it should probably be viewed more as a Kuhnian paradigm shift. The new constellation, in which more people are actively shaping technology, is different, but it may not take us any closer to machines whose design contributes to a more humanized world.

Democracy Versus Dystopia?

Feenberg's idea of the technical code represents a significant advance over earlier versions of constructivism. The principle that technical environments and their outputs are structured in a way that overdetermines them as technological and that this structuring constitutes a profound link to social power—that it *is* the impress of social power—is an important one. Without it, constructivism cannot account for the distinction that we all recognize (even if it occasionally breaks down in the application) between technical and nontechnical objects, still, less what that distinction signifies. It is important to see this as more than just a "semiotic hinge," linking our perceptions because of all that being technological signifies. Feenberg is surely correct to identify this as a crucial intersection of micro-contexts of design and use with wider questions of authority and social organization.

The central difficulty faced by the idea of technical codification, however, is that the agents who are constrained by the code and whose activities are informed by it at every step are the same as the ones who must produce the code, precisely by executing it in their practice.[47] The installation of new values in the heart of technology design turns on the appearance of networks in which new kinds of object and agent become thinkable. In other words, the kind of change Feenberg has in mind has, of necessity, an evental character. Much as innovations in art come from out of nowhere,[48] so technology invention sometimes inaugurates a new count, modifying reality itself as well as our ability to make sense of it.[49]

Feenberg's privileging of the concept of democracy as the central value of technical politics is underpinned by a "battle concept" inherited from the dystopian worldview of technocracy. There the problem of technology was understood in terms of experts using the authority of technology to dictate terms to technology users and, ultimately, imposing an unquestioning compliance on the rest of society. Technical politics is Feenberg's way of thinking through openings that ruptured this scenario, and democracy is an important value in that process.

However, with the change of cultural horizon involved in breaking with dystopia, our technological imaginary no longer equates it with the unquestionable authority of experts or even narrow efficiency. The diffusion of digital technologies has encouraged the development of diverse cultures of experimentation, dabbling, reconfiguring, sabotage, and so on. Much of this activity would have been unthinkable just a few decades ago. In this context, technology now exists in networks that extend into places not primarily associated with work and control. These networks produce and are produced by articulations of the technical code that are embedded in values like play and learning, rather than control or efficiency. Technology is as likely to be held equivalent to things that can be trusted, that are life-enhancing, individual, and even eccentric.

This is not to say that the technical code—the regime that governs our perception of artifacts and constrains what we think ourselves doing with them—has lost its connection with social domination. On the contrary, the entwinement of technology's network with those of social power seems to be as profound and pervasive in its consequences as ever. With the changed cultural horizon, however, the technical code is no longer subject to one or two kinds of articulation. Indeed, Feenberg himself makes this point when he writes that, "technical rationality consists of various loosely related dimensions with different social implications."[50] He does not draw the necessary conclusion for technical politics from this observation. If diverse rationalities are present in technology, then the competing values at stake are also already multiple.

Feenberg's case for social rationality as a complex and diverse phenomenon is compelling, but it stands in tension with the way he uses the concept most of the time. In effect, he needs to characterize the hegemonic technological rationality as monolithic in its articulation of the technical code to justify his version of critique, which defines itself as resistance to a singular power manifest in a recurrent form. What is the "technical illusion," after all, if not a pervasive form of "reification" structured around a particularly narrow construal of purposiveness and efficiency? Democratic technical politics is aimed at breaking this down. However, with the shift in constellations, from dystopia to where we are now, the technical illusion has evaporated to the extent that the internet, for instance, is a place where everyone dabbles in technology, or mobile phones become toys we all trust and incorporate into our lives.

These changes have succeeded in asserting difference, disrupting the dominant conception of technology and shifting the boundary of technique. As they have done so they have ushered in a re-politicization of technology design, which is now a normal part of the conversation for perhaps the majority of people. However, we have not seen a corresponding progressive change in society or culture. Democracy alone, it seems, cannot play the transformative role Feenberg assigns to it even if it is essential in creating space for other values.

Democratic reform of technology cannot be introduced in each of the three levels identified by critique unless it also arises in the other two: A design change is realistic if it is consistent with prevailing ideas of what technology is for and what it is. What technology is for is contingent on the kinds of artifacts that exist and the technological imaginary. The latter and our ideas of efficiency are in turn dependent on decisions people make at the scene of design. Perhaps the central difficulty faced by the idea of technical codification is that the agents who are constrained by the code and whose activities are informed by it at every step are the same as the ones who must produce it, precisely by executing it in their practice.

What seems to be needed here is a kind of "cut" through the prevailing socio-technical webs that transform what counts all at once. The ontological moment in technology is not an originary violence (primary instrumentalization), but more simply the appearance of new affordances that may follow upon the invention.[51] Feenberg points out that new technologies often threaten established authorities and social systems before they close around it again by regaining control of the technical code. This is the evental foundation of technical politics and it involves new openings where rival social logics can come to the fore and inaugurate new "counts," new orderings of the world.[52]

Democracy is one value that can be insinuated into discussions concerning who should be involved in making design decisions, and it makes sense to think that the more open design processes are the better. However, Feenberg's own theory tells us that what can be achieved this way must be limited. Technology retains dimensions that are incorrigibly paradigmatic rather than contingent on, or open to articulation—the boundary of the technique may shift, but it cannot be abolished.

Moreover, there is no easy association of democratization with the ethics (or aesthetics) of a superior civilizational model. Feenberg envisages "mutually supporting transformations"[53] as if this were a natural consequence of loosening hegemonic technological rationality's hold over the technical code. But the practice of eugenics in Sweden in the 1970s suggests that advanced democracy can go together with abusive medical technologies. Perhaps the underlying problem here is that the very conditions that make democratizing technology design thinkable also vitiate any notion of progress: in a pure democracy, the future has no direction aside from the one we give it.

It follows from this that the field of technology design in the changed constellation is actually more open to audacious assertions of nontechnical values than in the past. An important achievement of disabilities activism in the last few decades has been to raise the question of whether any given new device is properly inclusive. Part of the way we assess new objects is to ask whether they might create new inequalities. The fact that these discussions are already normal means that the possibilities for willed technology reform may actually be greater than Feenberg suggests. Democracy and critique cannot guarantee in advance, however, that contemporary struggles over technology design share a common direction or will end in the harmonious realization of human potential.

Notes

1. I would like to thank Andrew Feenberg for his typically generous comments on an earlier draft of this paper and for allowing me to read the early manuscript of his book, *Critical Constructivism*.
2. Andrew Feenberg, *Critical Theory of Technology* (Oxford: Oxford UP, 1991). *Transforming Technology* (Oxford: Oxford UP, 2002).
3. Herbert Marcuse, *One-Dimensional Man* (Boston: Beacon Press, 1964).
4. "...the technical sphere can be limited by nontechnical values, but not transformed by them." Feenberg *Transforming Technology* (Oxford: Oxford UP, 2002), 6.

5. Ernesto Laclau and Chantal Mouffe, *Hegemony and Socialist Strategy* (London: Verso, 1985).
6. Feenberg, *Between Reason and Experience* (Cambridge, Mass: MIT Press, 2010), 55.
7. Ibid., 57.
8. This point is emphasized in some constructivist studies: for example, "classifications and standards are *material* as well as symbolic," Geoffrey C. Bowker and Susan Leigh Star, *Sorting Things Out: Classification and its Consequences* (Cambridge, Mass: The MIT Press, 1999), 39
9. Feenberg, *Transforming Technology* (Oxford: Oxford UP, 2002), 63.
10. Feenberg, *Critical Theory of Technology,* 58–9.
11. Feenberg, *Between Reason and Experience,* xxiii.
12. "Technology must really work. But it is not merely because a device works that it is chosen for development over many other equally coherent configurations of technical elements." Feenberg, *Transforming Technology,* 79.
13. Bijker acknowledges that in most constructivist scholarship there is a "missing link" between these details and the question of power (Wiebe Bijker *Of Bicycles, Bakelites and Bulbs* (Cambridge, Mass: The MIT Press, 1997), 261). It is this link that Feenberg's idea of a technical code tries to provide.
14. Feenberg, *Transforming Technology,* 15.
15. Ibid., 20.
16. Feenberg, *Critical Theory of Technology,* 80.
17. Feenberg refers to a scale of meaningfulness for technological artifacts, ranging from "semantic impoverishment" to "richest object relations." Feenberg, *Between Reason and Experience,* 175.
18. Feenberg, *Critical Theory of Technology,* 81.
19. Michel Foucault, *The Order of Things* (New York: Vintage, 1994).
20. Ibid., 79.
21. In this it corresponds to what Bowker and Star describe as a classification system, forming "a juncture of social organization, moral order, and layers of technical integration" Bowker and Star, *Sorting Things Out,* 33.
22. Feenberg, *Between Reason and Experience,* 76.
23. Feenberg, *Critical Theory of Technology,* 14.
24. Laclau and Mouffe, *Hegemony and Socialist Strategy,* 93.
25. As in Ernesto Laclau, *Politics and Ideology in Marxist Theory* (London: New Left Books, 1979).
26. Laclau and Mouffe, *Hegemony and Socialist Strategy,* 183.
27. "The hegemonic form of politics only becomes dominant at the beginning of modern times, when the reproduction of the different social areas takes place by permanently changing conditions which constantly require the constitution of new systems of differences," Ibid., 138.
28. Ibid., 136.

29. Laclau and Mouffe write that in "industrial society there is a growing unification of the social terrain around the image of the mechanism" (Ibid., 36), which perhaps reflects the hold of a particular technological imaginary on their thought.
30. Feenberg, *Critical Theory of Technology,* 80.
31. Feenberg, *Alternative Modernity: The Technical Turn in Philosophy and Social Theory* (Berkeley: U of California Press, 1995), chap. 6.
32. Feenberg, *Between Reason and Experience,* 159.
33. Ibid., 166. For further discussion see Graeme Kirkpatrick "Formal bias and the normative critique of technology design," *Techné* 17(1) 2013.
34. Marcuse was famously accused by Habermas of espousing a dialogue with nature because he moved too quickly in developing his vision/fantasy of a radically different technology. See Jurgen Habermas, *Toward a Rational Society*, trans. J.J. Shapiro (Cambridge: Polity, 1989).
35. "Substantively biased decisions in the technological realm, where cool rationality ought to prevail, lead to avoidable inefficiencies…" Feenberg, *Between Reason and Experience*, 69.
36. Theodor W. Adorno, T.W., *Negative Dialectics*, trans. E.B. Ashton (London: Routledge, 1973).
37. Feenberg, *Transforming Technology,* 156.
38. Feenberg, *Between Reason and Experience,* 77.
39. Ibid., 80.
40. Ibid., 71.
41. This is characteristic of substantivist critique of technology, which from Heidegger to Borgmann bemoans technology as a source of superficial benefits bought at the price of profound impoverishment of our fundamental relationship with the world. See Martin Heidegger, *The Question Concerning Technology and Other Essays* (New York: Harper, 2013), and Albert Borgmann, *Technology and the Character of Contemporary Life. A Philosophical Inquiry* (Chicago: U Chicago Press, 1987).
42. Feenberg, *Critical Theory of Technology,* 61.
43. Feenberg, *Transforming Technology,* 33.
44. Feenberg, *Critical Theory of Technology,* 125.
45. Hence, Feenberg writes that "The meaning of modernity is at stake" in technical politics. *Transforming Technology,* 114.
46. Feenberg, *Between Reason and Experience,* 81.
47. Feenberg acknowledges that his technical politics contains a bootstrapping problem on this point.
48. Cf. Antoine Hennion, "Music and mediation: towards a new sociology of music" in M. Clayton, T. Herbert, R. Middleton, *The Cultural Study of Music: A Critical Introduction* (London: Routledge, 2003), 7.

49. "To enumerate is never an innocent operation; it involves major displacements of meaning," in Laclau and Mouffe, *Hegemony and Socialist Strategy*, 62–3.
50. Feenberg, *Critical Theory of Technology*, 178.
51. Bruno Latour points out: "...to the humble and honest work of making, they've [critical theorists] surreptitiously added a crazy hypothesis about the craftsman's domination of his oeuvre." Latour, *Rejoicing: Or the Torments of Religious Speech* (Cambridge: Polity, 2013), 142.
52. I have discussed this further in connection with the appearance of computer games in my "Early games production, gamer subjectivation andthe containment of the ludic imagination," in *Fans and Videogames: Histories, Fandom, Archives*, ed. M. Swalwell, A. Ndalianis, H. Stuckey (London: Routledge, 2017).
53. Feenberg, *Transforming Technology*, 27.

Bibliography

Adorno, Theodor W. *Negative Dialectics.* Translated by E.B. Ashton. London: Routledge & Kegan Paul, 1973.

Bijker, Wiebe E. *Of Bicycles, Bakelites and Bulbs.* London: MIT Press, 1997.

Borgmann, Albert. *Technology and the character of contemporary life: A philosophical inquiry.* London: University of Chicago Press, 1984.

Bowker, Geoffrey. C. and Susan L. Star. *Sorting Things Out: Classification and its consequences.* London: MIT Press, 1999.

Habermas, Jürgen. *Toward a Rational Society.* Translated by J. J. Shapiro. Cambridge: Polity, 1989.

Hennion, Antoine. "Music and mediation: towards a new sociology of music." In *The Cultural Study of Music: A critical introduction.* Edited by M. Clayton, T. Herbert, and R. Middleton. London: Routledge, 2003.

Feenberg, Andrew. *Between Reason and Experience.* London: MIT Press, 2010.

———. *Critical Theory of Technology.* Oxford: Oxford University Press, 1991.

Alternative Modernity: The technical turn in philosophy and social theory. Berkeley: University of California Press, 1995.

———. *Critical Constructivism*, forthcoming.

———. *Transforming Technology.* Oxford: Oxford University Press, 2002.

Laclau, Ernesto. *Politics and Ideology in Marxist Theory.* London: Verso, 1979.

Laclau, Ernesto and Chantal Mouffe. *Hegemony and Socialist Strategy: Towards a radical democratic politics.* London: Verso, 1985.

Latour, Bruno. *Rejoicing: or the torments of religious speech.* Cambridge: Cambridge University Press, 2013.

Marcuse, Herbert. *One-Dimensional Man.* Boston: Beacon Press, 1964.

Scharff, Robert and Val Dusek. *Philosophy of Technology: The Technological Condition.* Oxford: Blackwell, 2014.

CHAPTER 7

Andrew Feenberg's Ecological Modernism

Darrell P. Arnold

Introduction

Andrew Feenberg has long made important contributions to environmental thought. However, we nowhere find a comprehensive statement of what one might take to be his environmental position. In this contribution, pulling together ideas from his numerous writings, I outline a general statement of what we might take to be Feenberg's environmental philosophy to date, focusing on his calls for a green democracy. I then highlight some meta-issues in environmental philosophy and discuss considerations related to the global reach of the environmental problematic that complicate calls for a green democracy like Feenberg's. In short, here I argue that while Feenberg makes a very valuable contribution to the discussion of environmental philosophy, and green democracy in particular, he does not yet clearly enough address some important issues in green political and ethical theory. In particular, he does not carefully consider the demands for a democracy in a transnational framework; nor does he clearly enough reflect on how we secure voice for voiceless nature.

Feenberg's position largely aligns with what John Dryzek, following Peter Christoff and Marteen Hajer, has labeled a strong ecological

D.P. Arnold (✉)
St. Thomas University, Miami Gardens, FL, USA

D.P. Arnold and A. Michel (eds.), *Critical Theory and the Thought of Andrew Feenberg*, DOI 10.1007/978-3-319-57897-2_7

modernist perspective. Strong ecological modernists as described by these authors (1) call for broad-based changes to the institutional and economic structures of society so that they reflect environmental concerns; (2) call for open democratic and participatory decision-making procedures that foster communication about environmental issues; and (3) demonstrate a strong concern for the global character of the environmental movement.[1] While Feenberg clearly exemplifies the first two characteristics of this position, he needs greater clarity about the global character of the democratic movement that is needed to adequately address environmental policy concerns.

In Part 1 of this paper, I look at how Feenberg frames environmentalism within the philosophy of technology. I also examine his arguments for a needed paradigm shift in economics and its relation to his calls for a green democracy. In Feenberg's view, a green democracy is needed in order for us to get a handle both on the types of technologies we develop and on the type of economy we cultivate. In short, a green democracy is needed for us to achieve the type of ecological modernist industrial economy he thinks is possible. In Part 2 of the paper, I argue in ways complimentary to Feenberg's work but that move beyond it that there are good reasons for thinking that a green democracy is preferable to other forms of government and that we need to reflect in our legal system a stance that has more than a utilitarian view toward nature. Further, we need a multi-level and multi-dimensional approach to governance to address our technological and ecological concerns, since problems related to technologies and the environment have a global reach. In particular, official environmentally sensitive governmental mechanisms are needed at regional, state, and transnational levels that address our technological and ecological concerns. In addition, traditional government is not sufficient, but various extended forms of soft governance are also needed. Specifically, this means action in civil society, including global transnational movements and organizations, as well as the action of those moving to develop more sustainable and responsible businesses have a vital role to play in adequately addressing our long-range environmental concerns.

Part 1: Feenberg's Environmental Philosophy

Environmentalism and the Philosophy of Technology

Feenberg is in clear agreement with other philosophers of technology, like Don Ihde, that environmentalism is a major concern of philosophy of technology.[2] In his early *Critical Theory of Technology* Feenberg already acknowledges the role that the environmental crisis may play in bringing about a realization of the need for greater collective control over social and technological processes.[3] This view is only strengthened in his later writings.

In *Alternative Modernity*, for example, Feenberg emphasizes, among other things, the importance of the environmental movement in raising awareness of the value-laden character of technological decisions. For example, it has been effective at ensuring that questions of the design of nuclear plants and the storage of nuclear waste have become a part of public dialogue. Those in the environmental movement have been particularly skilled at using public forums (like Environmental Protection Agency hearings) to raise awareness that technological decisions are not value neutral but are connected to issues of human well-being.[4]

While Feenberg, like Heidegger and many other early philosophers of technology, accepts that technologies are not value neutral, and while he thinks that the environmental movement, in particular, is to be applauded for drawing attention to the value-laden character of technologies, Feenberg differs from Heidegger and numerous philosophers of technology and environmentalists who take an overwhelmingly negative stance toward technology. In Feenberg's view, technologies are not necessarily environmentally damaging. Hence Heidegger and others present us with a false dilemma when they presume that we have to choose between non-environmentally sound industrialism and environmentally sound preindustrial forms of life. We need to make no such choice. Instead, we can develop environmentally sensitive technologies.[5]

A fundamental problem, as Feenberg sees it, is that Heidegger and others who we might characterize, again in line with Dryzek, as eco-primitivists tend to erroneously think there is one line along which technological development occurs—specifically, that it moves in a deterministic manner toward a clearly defined goal of efficiency. What such thinkers fail to recognize is that technologies always develop along with human values—further, that values (good and/or bad) indeed become coded into the technologies. Standards of efficiency, too, are not value free. Rather they diverge in accordance with human values. The coding of our technologies can be for good or ill. In the best case, in Feenberg's view, a green democratic modernism would make strides towards ensuring that technological (including infrastructural) design embodies the human values that are aligned with our long-term interests, including the values of environmental concern. In sync with such values, it is still possible for a technology to efficiently meet goals. However, the standards of efficiency will not be exclusively related to profitability and they will have to consider long-term functionability. In particular, whether such a technology is sustainable should be basic to our understanding its long-term functionability and efficiency.

Feenberg roundly rejects the technological determinism that he views Heidegger and others as espousing—that is, the idea that "the existing industrial system is the only possible one," and that we either accept the results of industrialism or we revert to a pre-industrial system. By contrast, he asserts the possibility of different paths of industrialism, with a viable alternative form of industrialism being possible through the "incorporation of new values into industrial design."[6] In *Critical Theory of Technology*, he had already drawn on the work of Gilbert Simondon, arguing that it is necessary that our technological progress not only take into consideration nature as an externality but that in the creation of technological systems themselves, a concern for the effect on the human users of technology and on nature is needed from the outset.[7] In the kind of environmentally informed civilization that Feenberg imagines, in the "concretization" of technology—that is, the specific way that we encode technology with values—we will not promote technologies that continue the domination of nature but we will consider the integrity of nature as fundamental for the structure of any acceptable technology. In sum, environmental concerns will be fundamental to design questions. What is more, this will require a rethinking such that we see ourselves as constructing "synergistic totalities of natural, human and technical elements."[8]

Economic and Ecological Sustainability

In alignment with Feenberg's views that sustainable technological development is possible, he also argues that efficient economic development is compatible with environmentally sound policies. Precisely this is highlighted in "Incommensurable Paradigms: Values and the Environment," Chap. 3 of *Questioning Technology*.[9] In fleshing out his point, Feenberg contrasts his own approach with that of the early work of Paul Ehrlich on population growth as well as of Donella Meadows and other writers of *The Limits to Growth*, a work that famously calls for no-growth economic policies.[10] While these latter thinkers believe we have to abandon both industrial life and economic growth, Feenberg holds out hope in ecological modernization, with both developed technologies and economic viability.

Once an environmental concern is deeply established as a fundamental ethical concern, it will appear as odd to contrast economic development with environmental ethical concerns as it does to contrast economic well-being with the decision to be a prostitute or economic well-being with safety standards.[11] We have now accepted as basic that individuals not be forced into conditions of sex slavery or have to accept unsafe working conditions in order to make a living. Just as the economic well-being of some cannot now be viably argued to justify the subjugation of others, it should increasingly not be viable to argue that it is legitimate to compromise serious environmental concerns for economic well-being.

In developing his ideas, Feenberg draws on Thomas Kuhn's idea of a paradigm shift as it occurs in science. As Kuhn famously argued, such paradigm shifts not only occur against rational considerations but also involve (nonmoral) values such as the beauty of the models, or simplicity. Feenberg argues that something similar applies to developments of technologies and of economic growth. They do not merely roll out in accord with one best or efficient or rational standard. Rather, such developments occur against the backdrop of a set of varied concerns and with different social consequences.[12] Among the considerations of importance here are those of identity. In many cases, the developments that we undergo will depend on the type of culture or civilization that we want to have.[13] Neither technologies nor economies develop unhinged from values, but against the backdrop of particular values.

The technological revolution, and the accompanying economic revolution that Feenberg expects in light of the new environmental concern, may look irrational from a present perspective of economic efficiency. But we are in the midst of a paradigm shift that defines *civilizational values* as discussed. In the future, the now existent view of economic efficiency will be viewed as deeply flawed and itself irrational. Revolutions, as Feenberg notes, judge the present in reference to an imagined future. This is precisely what environmentalism is now doing.[14] We are living in the midst of an environmental revolution. Feenberg projects that future generations "will accept environmentalism as a self-evident advance. Just as images of Dickens of the bootblack factory testify to the backwardness of his society, so will the images of asthmatic children in smog-ridden cities appear to those who come after us."[15] Options viewed now as live will in the future be viewed as "dead options."[16]

The green eco-industrialism that is possible will require a rejection of the models of efficiency that have been dominant in Capitalism, traditional Socialism, and Communism. In *Critical Theory* Feenberg acknowledges that, in contrast to early forms of Communism, which affirmed the same model of efficiency and productivity as Capitalism and only maintained that Communism would be more efficient than Capitalism, later forms of progressive politics, especially associated with the present green movements, have more blatantly challenged the predominant model of efficiency and productivity. For some time they have argued that Socialism might be less productive than Capitalism but more fulfilling. Or they are less productive of certain goods.

One of the largest problems that Feenberg sees with our existing economic model is that the standard for measuring success is tied to GNP, as the simplest measure. In contrast to this, he argues for readjusting the measure for efficiency away from a pure economic standard and toward a more subtle and richer analysis of human well-being that makes reference to the educational level of a population, its health, its ability to protect the weak, the satisfaction of a population with the labor practices, and the quality of the environment.[17] His measures point in the direction of the UN greatest happiness index, the Human Development Index, or of considerations of the OECD on what constitutes a good life.[18] The point is that such measures still value modernist efficiency. But they measure that efficiency differently than a model that looks only at GNP. Feenberg is critical of a tendency of many environmentalists to reject the value of efficiency altogether.[19] What we need instead is a new paradigm for efficiency.[20]

While this paradigm shift is needed, Feenberg is not merely advocating a need for a shift in attitudes. He is ultimately aiming at the structural transformation of our economic and political systems. In *Between Reason and Experience* and more recently "The Liberation of Nature" Feenberg emphasizes the connection of his views about the need to redefine efficiency with ideas of Marcuse. Like Marcuse, he thinks that the attempt to redefine efficiency in connection with "overall human well-being" is difficult to manage under the constraints of capitalism and under our current democracy.[21]

On Greening Democracy

In *Between Reason and Experience*, Feenberg rejects two views widespread in environmental philosophy: first, that we will need some spiritual shift in consciousness, as imagined by Paul Ehrlich—and, we might add, deep ecologists—and, second, that we will need a benign environmental dictator to establish a green polity, such as Heilbroner suggests[22] but, also for example, as Hans Jonas had argued. As Feenberg notes: "It's difficult to decide which of those solutions is more improbable."[23] His own solution aims at shifts in technological design, the economy, and our form of democracy. Such changes for their part do also demand the transformation of culture.

A green democracy would operate with an understanding that technological designs and the laws about the design of technology are inherently ethically and politically laden. It would come to reflect the concerns for the environment both in general laws about how things can be produced but also in laws about what is produced. Various views of what this democracy would look like are touched upon in Feenberg's writings on early figures of importance in the environmental debate. They are described more fully in texts on the democratization of technology. But the writings on what these democratic systems will look like lack specificity. Further, we can see, it is questionable to what extent our present democratic system appears anywhere close to ready to achieve what Feenberg has in mind. Like his teacher Marcuse, Feenberg, of course, recognizes the challenges we confront in our present democracy. In later work, in line with the work of Adorno and Marcuse, Feenberg emphasizes the need for a shift in aesthetic sensibilities. Such a democracy that we will need for both the right kind of technologies and the right kind of economy requires a differently aesthetically sensitized citizenry. A look at the Scandinavian countries, which have done the best at developing the type of green policies Feenbeg has in mind, corroborate Feenberg's view that a "spiritual transformation" such as is called for by Ehrlich or Deep Ecologists is hardly necessary for the green democracy. Yet one would still be surprised if religions and forms of spirituality that exist at the time of an ecological paradigm shift did not also reflect the general aesthetic education, or are not a part of such an aesthetic education.

In any case, Feenberg rejects a traditional Marxist perspective, accepted, for example, in the early work by Barry Commoner, that an adequate change to social structures can occur without a needed change to culture. Following in the footsteps of Marcuse, Feenberg thinks that cultural changes work together with structural changes. Cultural changes are needed to get the kind of democracy he thinks needed.[24]

Revisiting the Frankfurt School—On the Need for a Structural and Cultural Transformation

The most recent of Feenberg's writings devoted to the environment is "The Liberation of Nature," In this text, Feenberg outlines what he believes to be a fundamental contribution of the early Frankfurt school and especially Adorno and Marcuse to the environmental thematic in a way that might be useful today. The main gist is that their work can contribute to our understanding of what is needed to create cultural conditions that make possible a shift in our technological, economic, and political systems.[25]

In Feenberg's view, the main contribution that Frankfurt School critical theory can make to environmental philosophy is related to this aesthetic reorientation of our engagement with the natural world. It suggests a way to overcome the antinomies of subject and object and of facts and values. "On this basis it conceives a utopian vision of a life-affirming form of individuality and the corresponding politics and society."[26] His main point is that an aesthetic sensibility aligned with Marcuse's (and Adorno's) can serve us in shaping the world in ways that allow a freer and more commodious life for humans and that at the same time frees nature from the domineering tendencies of humans, as formed under existing systems, where we are driven, not by aesthetic considerations, but by an instrumentalizing framework that sees in nature nothing other than a resource to be exploited in any way that leads to a perceived economic benefit. Both our political and economic structures as well as the development of our technologies need to be informed by such broader aesthetic considerations.

As Feenberg understands this, rather than perceiving nature as entirely separate from value and as lacking potentialities, we need to recognize, perhaps more clearly than Marcuse could, that it has long been shaped by human value and by human views of its potentials. The dominant existing approach to nature, which denies it essence and potential, ultimately paves the way for degrading nature. A renewed approach to nature, informed by aesthetic considerations, would mean that we would be more likely to shape nature in a manner beneficial to its own flourishing and purposes that attune with a broader understanding of well-being (in which having a flourishing environment is viewed as vital to having a good life more generally). Feenberg here uses "aesthetic" in a broader Kantian and Marcusean sense as applying to our appropriation of sensuous reality more generally. The subject is involved in the mental production of objects. However, in line with the Frankfurt School, Feenberg emphasizes that we not only are involved in the mental formation of the objects of nature but also in the physical production of the objects in the natural world. An aesthetic understanding ought to influence not only how we perceive the world but also how we shape it. As we become more cognizant of the human role in the conceptualization of the natural world and in the very shaping of the natural world itself through our labor (and broader action), we also come to see that values are at play in the (mental and physical) production of the natural world. In accord with Feenberg's Marxist view of the antinomies, such an understanding of how values are already involved in the mental appropriation and the physical shaping of the world around us allows us to overcome the antinomy of subject and object since it brings to light that subjects have long shaped objects. It also shows a resolution to the antinomy of fact and value since it unveils ways that values have long shaped facts.

Feenberg does emphasize that failing to regulate on environmental questions is not the solution: "The conservative antiregulatory approach is based on an illusion. It forgets that the design process always already incorporates standards of safety and environmental compatibility."[27] In fact, noneconomic values and economic values intersect in design codes.[28] Further, "technical design standards define major portions of the social environment, such as urban and built spaces, workplaces, medical activities and expectations, life patterns, and so on."[29] Such standards influence an entire way of life and "the cultural framework of the economy."[30] Because the tendency in capitalism is to reduce the plethora of goods to GNP or some other easily achieved monetary measure and to contrast these to "economic efficiency," Feenberg's argues that a democratic socialist system will better enable us to enframe the economy in terms of broader goods of a whole life, which includes environmental goods.[31]

In sum: Feenberg agrees with Marcuse that in a future ecological modernist society we would not return to premodern crafts. Rather, the way in which we would shape our technologies and our social world would more explicitly incorporate life-affirming values, including environmental values. Instead of being so fully dominated our considerations by narrow interests in economic growth, we will incorporate broader human and natural needs that are part of Adorno's and especially Marcuse's general aesthetic considerations. A green democracy is ultimately needed to ensure these developments.

Part 2

Systemic Issues in Environmental Philosophy and Questions of a Global Green Democracy

While this presentation of Feenberg's contribution to environmental philosophy does not offer all the details of his position or highlight where he stands on some particular issues, such as overpopulation, which he has written about, it provides the main outlines of his work. If anything characterizes Feenberg's writings on environmentalism it is the call for democratization so that we can gain greater collective control over both our technologies and our economy, both of which are often falsely viewed as value free.

While Feenberg's views are very useful as far as they go, Feenberg's major reflections on environmentalism do not deal with some issues of great importance to debates in environmental political theory. Further his views of democracy remain at a very abstract level and do not deal adequately with issues in green political theory that are important given the global character of the environmental problems and the typically regional and statist character of our political bodies. In this section of the paper, I will address some systemic issues in environmental philosophy which are not addressed by Feenberg but in ways that I think largely complement Feenberg's perspective. I will also offer reflections on issues of global governance that indicate possibilities for augmenting Feenberg's perspective.

Expert Versus Participatory Decision-Making

One major systemic issue is that Feenberg seems to conflate democratically agreed-upon outcomes with environmentally sound ones. Yet, as Robert Goodin long ago pointed out, democracy describes a procedure, but what is important for environmentalism are results.[32] Democracy is only as good for the environmental problematic as its results prove. This is not to suggest that an eco-elitist decision-making is better for our predicament, but this point deserves to be addressed in some greater detail.

While it is easy to find agreement with Feenberg that the strong ecological authoritarian position that Heilbronner and others espoused is absurd, since we could have little guarantee that any authoritarian government would remain green and would have virtually no means to redress such a government should it fail to be, it is less clear that the position of those who Dryzek characterizes as "administrative rationalist" is without merit. Regarding decision-making, administrative rationalists propose technocratic policy procedures, led by scientific and governmental elites. According to administrative rationalism, a representative democracy with similarities to those in the industrialized West and that relies on political and scientific experts will be more practicable and ultimately provide for more environmentally friendly results than the type of participatory democracy proposed by Feenberg. For example, this view seems to be embraced by Anthony Giddens.[33]

Proponents of administrative rationalism emphasize certain weaknesses of the general citizenry that prohibit it from making good decisions about issues of science and technology that are relevant to the green agenda. A couple of the more prominent reasons follow: (1) Non-experts have little interest in science and technology and are not likely to develop such interests, so as a practical concern it is not likely that they will be successful at advising on scientific or technical developments. (2) Non-experts are largely susceptible to consciousness distorting biases, such as the availability heuristic that leads people to consider recent major examples of some phenomenon and weight these as more important than they should when making decisions. It is due to the availability heuristic, for example, that the sale of tornado insurance increases after a tornado, even though there is no spike in the need and a sober assessment might show that some other form of insurance is a more rational investment. A group of other heuristics that distort decision-making can be added to this one. Experts, who know their fields, are less likely to be taken in by such irrational forces.[34]

By contrast, proponents of more participatory democratic solutions can argue the following: (1) Non-experts are not as subject to professional biases or fears of damage to their reputation if they change their position as those who are experts in sciences and technology. (2) The involvement of non-experts has some social values: (a) The involvement of non-experts plays a valuable educational function since the attempt to democratize questions of science and technology leads citizens to become more informed about these areas than they would be otherwise. (b) If the procedures are viewed as participatory and legitimate, people have also been shown to be more accepting of the social risks produced by such decisions, since they were involved in the process of assessing them and decisions about how to address them. (3) In sync with a view that autonomy requires that "what affects all should be decided by all" a system of deliberative democratic decision-making fulfills requirements of procedural justice that align with individual autonomy.

While I can only provide a bare-bones defense of the value of deliberative democratic procedures over those of administrative rationalism, I do think this begins to show why more deliberative democratic and participatory procedures are preferable to expert oriented ones. In the type of deliberative democratic system proposed here, experts will still be fundamental to the processes, but in such a system an attempt is made at every step to involve the public as much as possible in the decision-making process. The involvement does not mean that citizens will make decisions about the scientific merits of various developments, but that they will be actively engaged, reflecting value concerns. This aligns in general outline with a diverse body of work on science and values, from Phillip Kitcher's views of "well-ordered science" to the arguments of various feminist philosophers of science like Sandra Harding and Helen Longino.[35]

Still, the most recent election cycles in the United States show that at least the USA is far indeed from having a citizenry that has the aesthetic and scientific sensibility that is needed for a green democracy. Recent elections of George W. Bush and Donald Trump to the office of President of the United States (the former who long questioned the human contribution to climate change and the latter who has deemed it a hoax perpetuated by the Chinese) appear to signal that a civilizational paradigm shift is still a somewhat distant dream. Unfortunately, we remain not far removed from the concerns at the time of Marcuse's writing of *One-Dimensional Man* that while a more participatory democracy may be our greatest hope, such a democracy (now with a green agenda) will only really be possible with a citizenry that has undergone what Marcuse, following Schiller, viewed as an aesthetic education.

Questions of Anthropocentrism

Another fundamental question of environmental philosophy and environmental political theory, in particular, concerns whether or not anthropocentric standards for our decision-making are sufficient to ensure a robust green democracy.[36] This debate, too, is complex. Among the diverse metaethical stances of importance in environmental philosophy are anthropocentric positions, along with zoocentric and bio- or earth-centric ones. Anthropocentric ethicists typically argue that we can get everything we need for a robust environmental ethics just in light of human concerns. As long as we have a reason to be ethically concerned about future generations of humans, we have a reason to take care of the environment, since the well-being of those future generations of humans is dependent on having a well-preserved environment. Zoocentrists expand the circle of our moral concern, maintaining that our ethical and political concerns should include both human and other animals (typically providing a scale of animals of concern depending on their ability to feel pain). Peter Singer is the best-known proponent of this position. Such zoocentrists typically argue not only that the suffering of animals means that we should consider them in our moral deliberations, but also that doing so will result in more robust environmental policies. If their good is viewed as merely instrumental to the good of humans, it is argued we will be too ready to sacrifice their good for ours. Biocentric or earth-centric positions extend moral consideration even beyond the animal kingdom. The concern of bio-or earth-centrists is not only the pain that certain creatures can feel but the well-being of biological or even non-biological nature more generally. Besides offering metaethical arguments that the care for the biome—or the land or earth more generally—is the best rationally defensible position, these ethicists argue that only a robust bio- or earth ethics will best ensure that we will not encroach on the well-being of nature for some short-term and non-essential interests of humans (or animals). Sometimes the positions of various zoocentrists or bio- or earth-centrists will elevate nonhuman ethical concerns to the same level as human ethical concerns. Each of these debates has a level of sophistication that cannot be taken up here. It is clear that whether or not one or the other of these meta-views will render a more robust environmental policy than the other is an argument that can really only be adequately addressed ex-post-facto. Some evidence that one or the other of this meta-position might best allow us to address environmental concerns would be available after various governments enacted legislation with varying underlying moral and legal arguments. But all of these arguments have the difficulty of often involving us in speculations on counterfactual assumptions, and it remains unclear whether even communication free of distortions such as Habermasians would aim at would show one or the other position to be clearly more rationally forceful.

Despite these difficulties, I think it is informative to quickly examine, at least for plausibility, the positions of two critical theorists working in an area sometimes characterized as "critical political ecology" and who argue that we would do well to move beyond an exclusive focus on anthropocentric interests.[37] Both Robyn Eckersley and John Dryzek do work in green political theory, attempting to extend the work of the Frankfurt School in directions relevant for the ecological crisis, and attempting to compensate for what Eckersley has called "the failed promise of critical theory"[38] to address environmental and technological concerns. Both thinkers align more with the early research program of Horkheimer and the work of Marcuse rather than the more pessimistic later work of Horkheimer and Adorno. Both also incorporate and ameliorate some of the ideas developed in Habermas's critical theory, trying to extend the Habermasian model to allow for representation of voiceless nature and future generations in our political systems.

While in her early work Eckersley focused on *ecocentric* greener consciousness, in more recent work, she seems to have struck a more pluralistic tone in writing more formally directed toward formal statist political structures.[39] Her 2004 book, *The Green State*, does not focus on ecocentrism, but she still speaks in various places of an "ecological democracy" with policies "that go beyond a purely instrumental posture toward the nonhuman world."[40] Dryzek, whose position Eckersley has to some extent moved toward, also wants to surpass a mere anthropocentric and certainly an instrumentalist understanding of nature. He has suggested we take more literally than James Lovelock perhaps intended his concept of a "non-anthropocentric democracy."[41] But he maintains that this does not imply an ecocentric agenda, presumably of putting other natural interests on equal footing with human interests. Instead, he advocates a green politics that transcends the "anthropocentric" or "ecocentric" divide while still providing a "more egalitarian exchange at the human/natural boundary."[42] Since nature has agency—though not subjectivity—we need to "listen to signals emanating from the natural world"[43] and accord them some of the respect we accord to human agents who can speak for themselves.[44] In a similar tone, Eckersley references the need to govern for common "communities of fate," seeming to acknowledge at least in some general sense the early sentiment like that of Hans Jonas "the common destiny of man and nature," which commands a respect for the integrity of nature "above the utilitarian aspect."[45] What both Eckersley and Dryzek emphasize is the need for democratic reforms that allow as much as possible the participation from all of the affected. Since future generations and voiceless nature cannot speak for themselves, for both Dryzek and Eckersley there will be stand-ins for them in the political processes. This entails a non-anthropocentric, but not necessarily a zoocentric, or eco- or earth-centric position. It does involve expanding the circle regarding what is viewed as moral considerable and consciously attempting to give voice to those concerns (but without elevating the concerns of nonhuman nature necessarily to the level of human concerns).

Feenberg's position generally seems anthropocentric, though some of his discussion of Simondon, noted earlier in this essay, indicates the possibility of moving beyond a merely instrumental relationship with nature that aligns with the intent of Eckersley and Dryzek. How justified are concerns that such a position may well fail to secure as robust of protections for nature as we need? We probably cannot know. In any case, Feenberg has acknowledged that environmental concerns need to be reflected in technologies. Still, he has said little about how this would be secured in the legal or R&D systems that are in place. Dryzek and Eckersley have suggested citizen juries and boards with representatives for animals and future generations. They have also suggested juries and other boards that represent such interests, for example, from those outside of the nation-state enacting the particular policies in question but not included in the political system as voters, despite that they are affected strongly by the decisions of that nation-state. Might it be recommendable to have such boards in both political bodies and R&D meetings of companies of a certain size to secure as best as possible that the needs of nature are taken into consideration in our designs? The legal system would need to provide guarantees that nature is dealt with "above the utilitarian aspect" if only to secure that short-term interests do not trump long-term ones.

Questions of a Global Green Democracy

Besides the question of which forms of life are represented formally in government as well as civil society and business decision-making, other questions regarding democracy are very relevant to any green democratic theory. First, how can we more adequately democratize our technologies and economies in governance systems that are largely national and regional, even though the effects of technological and economic developments of national and regional bodies are global? Short of a Global Parliament (which can hardly be seen as emerging any time soon), how might our statist democratic systems integrate global concerns and provide a global voice for those affected who live outside the boundaries of the democratic state in question? The above discussion pointed to the possibility, for example, of formal representative bodies entrusted with an environmental agenda.

However, deliberative democracy understands democracy to consist of more than voting procedures and the work of formal governmental bodies and to include life in civil society. One has increasingly good reason to think corporate policies and conscientious consumption movements are important here as well. Clearly, civil society organizations exert pressure on various national governments around the world to enforce environmental standards. They can be used to leverage for corporate laws that require formal consultation with representatives from government or NGOs that represent the interests of nature. Or they can pressure companies directly, for example, through boycotts, to implement more environmentally sustainable practices. We of course already see the value of some such transnational movements in groups like Green Peace or Earthfirst! that, sometimes through their radical work, raise consciousness about environmental issues and can thus influence governmental and business policies. These and similar groups also play an important role in the identity formation of the greater public so that it more seriously considers global ecological concerns. An added value of the transnational character of such organizations is that it provides them with conduits for sharing information from marginalized countries with those in more powerful ones and for generating information that facilitates global identity and global action. Though we may be a long way from a Global Parliament, for those of us who see this as ultimately valuable, it is worth noting that work at creating a global identity is a necessary part of the cultural education needed for a possible later structural transformation. Further, it helps sensitize those involved in such transnational or global networks to important issues that they can respond to now through either market, social, or existing governmental channels.

In sum, the work of transnational civil society is important for helping to sensitize a citizenry to global concerns so that they work for formal governmental mechanisms at their regional and national and transnational levels that reflect a concern for global environmental (and technological) issues. Such global identity formation work is necessary for creating the cultural conditions needed for a possible later Global Parliament.

Can We Do More Than Tame Capitalism?

Milton Friedman famously said, "we are all Keynesians now."[46] While he was hardly speaking for the Left, we do see movements in that direction from numerous voices, including Habermas, who sees us as able to do little else than "tame" capitalism. Against the background question of what a transformed green economy would really look like in detail, I would like to highlight the question of the role that corporate social responsibility and conscientious consumption might play in the process of the cultural and structural transformation of society. The subject is borderline taboo for critical theory since the ultimate goal of critical theory generally is to fundamentally change the economic system. Here, too, though, it seems important to face the fact that no such change in the economic system appears to be pending. So what can we do in the interim? And indeed, though Feenberg calls for a green economy, the details that he has in mind for this are not clear. Even in a green economy, I am supposing that some individuals would own shops and that there would be considerable freedom in purchasing decisions, and that such decision could continue to have an effect on the organizational policies of the existing enterprises. So the discussion of organizational culture and of the importance of everyday personal decisions, even in our purchasing decisions, is certainly important now, and it will likely retain importance. I would like to suggest, for example, that the support of organic products or cradle-to-cradle production is relevant as part of the general cultural transformation of society. In fact, a green society will not come about merely because of changes to governmental policy. Rather, any such changes, especially if they are to be enacted by a green citizenry, will have to be internalized by that citizenry and lived out in everyday patterns of consumption. The green consumption and the green production movements are both important parts of the creation of a sustainable culture and economy. They both help in the formation of identities needed for a well-functioning green democracy and the greener technologies that emerge in part under the pressures of the conscientious consumption and green production movements in many cases will be the same green technologies that are needed in any economic system.

Conclusion

Andrew Feenberg has made a positive contribution to environmental philosophy especially with his acute reflections on the views of technology and economics that underline certain key positions in the environmental philosophy debate and by highlighting the need for a green economy and green democracy. However, his reflections on the details of the economy and democracy do not pay enough attention to the global character of the economic and political orders that are so important in technology development; nor does he focus enough on how to institute environmentally sound policies in those arenas. My hope is that the reflections in Part 2 of this paper, among other things pointing to work in critical political ecology, show directions fruitful for participatory democrats like Feenberg.

Unfortunately, we are in fact still showing ourselves quite unprepared to confront the colossally difficult issues related to the global reach of our technologies and the global character of the environmental crisis. Among the most important issues is that there is a problem with the existing statist governmental structures given the global character of so many of our technological and environmental concerns. Nonetheless, the cultural changes that can be leveraged to help improve our current statist and corporate systems are also the ones that can help create a more global cultural identity that is a prerequisite for an eventual Global Parliament and environmentally sensitive economy. From where we find ourselves, I am less optimistic than Feenberg that we will, in fact, manage the cultural paradigm shift and move toward the needed green democracy and green economy. Nonetheless, action in the existing formal governments, action in global civil society and action aimed at reforming business organizational culture and ultimately the economic system so that they are more sustainable, remain our present best hope.

Notes

1. John Dryzek, *The Politics of the Earth* (Oxford: Oxford UP, 2013), 173 ff.
2. Don Ihde, *Philosophy of Technology. An Introduction* (New York: Paragon House), 119–28, 193.
3. Andrew Feenberg, *Critical Theory of Technology* (Oxford: Oxford UP, 1991), 17.
4. Andrew Feenberg, A*lternative Modernity: The Technical Turn in Philosophy and Social Theory* (Los Angeles: University of California Press, 1995), 38ff.

5. Ibid., 35.
6. Feenberg, *Critical Theory of Technology*, 125.
7. Ibid., 193ff.
8. Ibid., 195.
9. Andrew Feenberg, *Between Reason and Experience: Essays in Technology and Modernity* (Cambridge, Mass: MIT Press, 2010), 31.
10. Ibid, 31ff.
11. Ibid., 35.
12. Ibid., 38.
13. Ibid., 42.
14. Ibid., 45.
15. Ibid., 43.
16. Ibid., 44.
17. Feenberg, *Critical Theory of Technology*, 126.
18. The Human Development Index of the UN Development Program is another such measure. It specifically looks at the three categories of (1) "a long and healthy life," (2) "knowledge," and (3) "a decent standard of living." See, for example, http://hdr.undp.org/en/content/human-development-index-hdi. The UN Development Program, for its part, acknowledges that this is an incomplete list of issues of what constitutes a good life, but sees this as a model that is easier to use than others and that does provide for a more adequate understanding of well-being than GNP alone. According to its 2015 report, Norway ranks number 1, the US number 8 (available at http://hdr.undp.org/en/2015-report). On various rankings that include more complex listings, the US ranks lower still. One such standard is the world happiness report, which includes an evaluation of GDP, social support, healthy life expectancy, freedom, generosity, and lack of corruption. In the 2016 ranking the US was 13, Denmark 1. (See http://worldhappiness.report/wp-content/uploads/sites/2/2016/03/HR-V1_web.pdf). Further measures, with more complex rankings are proliferating. The homepage of the OECD, like the websites noted above, facilitates thinking along the lines proposed by Feenberg in this early work, asking for participants to indicate what they view as the most important components of a good life. It's own first suggestions include a weighting of housing, income, jobs, community, education, environment, civic engagement, health, life satisfaction, safety, and work-life balance.
19. Feenberg, *Critical Theory of Technology*, 124ff. Cp. 158.
20. The focus on the need for such a new standard is found throughout Feenberg's work. Already in *Critical Theory of Technology* he argues we are at the cusp of a "civilizational change," where what it means to be human is redefined. Here and in later work, he sees especially the environmental

movement as moving us toward a redefinition of human life, human value, and economic viability (*Critical Theory of Technology*, 146ff.).

21. See "The Liberation of Nature," 19ff. Feenberg links such movements to the early work of William Morris, who contrasted "useless work" with "useless toil" when arguing to restore traditional crafts and to affirm traditional community (*Critical Theory of Technology*, 124–5). He also joins this to a view of politics as we will see later and views Mumford's contrast between "democratic technics" and "authoritarian technics" as fitting into this tradition, and connects it to Amory Lovin's differentiation between "soft" and "hard" technologies. This is the direction in which Feenberg thinks we must move.
22. Cp. Feenberg, *Questioning Technology* (New York: Routledge, 1999), 93ff.
23. Ibid., 94.
24. He notes that in Commoner's later work that he does see the value in personal commitments and volunteerism related to the environment, especially for the way it contributes to cultural changes (Ibid., 67). Yet, Commoner's main thrust underemphasizes cultural changes. As Feenberg argues, "He was so busy with his polemic against individualistic environmentalism that he rejected all concern with culture, which, he seemed to fear, would lead back to lifestyle politics" (Ibid., 65). In Feenberg's view we need to surpass Commoner's naïve faith that the labor parties will address the environmental cause and bring about the needed social change to inaugurate a new environmentally sound political age. We need a transformation of culture that leads to the introduction of conditions that make possible the needed transformation of social structures so that we can initiate a new green politics and a new greener understanding of economic well-being and new greener technologies. Cultural and structural change need to occur hand in hand. To say the least, however, achieving this seems a mammoth task.
25. In Feenberg's view, the main contribution that Frankfurt School critical theory can make to environmental philosophy is related to this aesthetic reorientation of our engagement with the natural world. It suggests a way to overcome the antinomies of subject and object and of facts and values. "On this basis it conceives a utopian vision of a life-affirming form of individuality and the corresponding politics and society" ("The Liberation of Nature," included as the Afterward in *Critical Ecologies: the Frankfurt School and Contemporary Environmental Crises*, ed. Andrew Biro (Toronto: University of Toronto Press, 2011), here: 344).
26. Ibid., 344.
27. Ibid., 22.
28. Ibid., 23.
29. Ibid.

30. Ibid.
31. Ibid., 28.
32. Robert Goodin, *Green Political Theory* (Cambridge: Polity, 1992), 168.
33. See Anthony Giddens, *The Politics of Climate Change* (Cambridge: Polity, 2011).
34. For a detailed treatment of the role of problematic heuristics that can interfere with rational decision-making, see, for example, Cass Sunstein and Richard Thaler's *Nudge Improving Decisions about Health, Wealth, and Happiness* (New York: Penguin, 2009). For Sunstein's treatment of such heuristics and risk, see *Risk and Reason* (Cambridge: Cambridge UP, 2002). Various ideas from the next paragraph are supported in that book.
35. For Kitcher's views on well-ordered science, see *Science, Truth and Democracy* (Oxford UP, 2003) and *Science in a Democratic Society* (Amherst, NY: Prometheus Books, 2011). I offer a succinct treatment of Longino and Harding in "Political Epistemology: Hegel and Critical Feminist Philosophy of Science," *Hegel-Jahrbuch* 10 (2008): 319–324.
36. This for its part is tied into a group of questions in environmental philosophy, addressed, for example, in an exemplary manner by Holmes Rolston III, about whether nature has intrinsic value at all, why, etc.
37. For an overview of some of this material see my "The Greening of Democracy," in *From Ego to Eco: Imagining Ecocriticism in Literature, Film and Philosophy*, eds. Tina Pusse and Sabine Mueller, forthcoming (Amsterdam: Rodopi).
38. Robyn Eckersley, *Environmentalism and Political Theory. Toward an Ecocentric Approach* (Albany, NY: State University of New York Press, 1992), 95.
39. Robyn Eckersley, "Ecocentric Discourses: Problems and Future Propects for Nature Advocacy," in John Dryzek and David Schlosberg (eds.), *Debating the Earth* (Oxford: Oxford U P, 2004), 364–382. Here Eckersley argues that the ecocentric discourse has an educative role in civil society and that it helps set the tone for debate in political systems.
40. Eckersley, *The Green State* (Cambridge, Mass: MIT Press, 2004), 251.
41. Dryzek, *Deliberative Democracy and Beyond*, 151f.
42. Ibid., 146.
43. Ibid., 148.
44. Though Steven Vogel is right to contend we cannot speak for nature in itself, but only nature as we have constituted it, even if our information about nature's needs is imperfect, we still must accord a dialogue about those needs a place within our political systems. See Steven Vogel, *Against Nature: The Concept of Nature in Critical Theory* (Albany: State University of New York Press, 1995). See also Vogel's "Habermas

and the Ethics of Nature," in *The Ecological Community* (New York: Routledge, 1997), 175–92.

45. Hans Jonas, *The Imperative of Responsibility* (Chicago: University of Chicago Press, 1984), 137.
46. See Milton Friedman, "Letter: Friedman & Keynes," Time, Feb. 4, 1966: In the famed passage, Friedman of course said we are only Keynsian in one respect and that in another respect "no one is a Keynesian any longer."

Bibliography

Arnold, Darrell. Political Epistemology: Hegel and Critical Feminist Philosophy of Science. *Hegel-Jahrbuch* 10 (2008): 319–324.

———. The Greening of Democracy. In *From Ego to Eco: Imagining Ecocriticism in Literature, Film and Philosophy*, edited by Tina Pusse and Sabine Mueller. Amsterdam: Rodopi. Forthcoming.

Dryzek, John. Political and Ecological Communication. In *Ecology and Democracy*, edited by Freya Mathews. London: Frank Cass, 1996.

———. *Deliberative Democracy and Beyond*, Oxford: Oxford UP, 2000.

———. *The Politics of the Earth, Environmental Discourses.* Oxford: Oxford UP, 2005.

———. Fo*undations and Frontiers of Deliberative Governance.* Oxford: Oxford UP, 2013.

Dryzek, John and David Schlosberg, Eds. *Debating the Earth.* Oxford: Oxford UP, 2004.

Eckersley, Robyn. *Environmentalism and Political Theory. Toward an Ecocentric Approach.* Albany, NY: State University of New York Press, 1992.

———. *The Green State.* Cambridge, MA: MIT Press, 2004.

Feenberg, Andrew. *Critical Theory of Technology.* New York: Oxford UP, 1991.

———. *Alternative Modernity.* Los Angeles: University of California Press, 1995.

———. *Questioning Technology.* New York: Routledge, 1999.

———. *Transforming Technology: A Critical Theory Revisited.* Oxford: Oxford UP, 2002.

———. *Heidegger and Marcuse: The Catastrophe and Redemption of History.* New York: Routledge, 2005.

———. *Between Reason and Experience: Essays in Technology and Modernity.* Cambridge, MA: The MIT Press, 2010.

———. *The Philosophy of Praxis: Marx, Lukács and the Frankfurt School.* New York: Verso, 2014.

Heidegger, Martin. The Question Concerning Technology. In *Martin Heidegger: The Basic Writings*, edited by David Farrell Krell. San Francisco: HarperSanFrancisco, 1993[1963].

Hellliwell, John, Richard Layard and Jeffery Sachs, Eds. *The World Happiness Report 2016*. Accessed Jan. 2, 2017 at http://worldhappiness.report/wp-content/uploads/sites/2/2016/03/HR-V1_web.pdf.

Honneth, Axel. 2009. *Pathologies of Reason: On the Legacy of Critical Theory*. Translated by James Ingram. New York: Columbia UP.

Ihde, Don. *Philosophy of Technology. An Introduction*. New York: Paragon House, 1999.

Jonas, Hans. *Das Prinzip Verantwortung*. Frankfurt a.M.: Suhrkamp. 1979/2003.

——— *The Imperative of Responsibility: In Search of an Ethics for the Technological Age*. Chicago: University of Chicago Press, 1984.

Light, Andrew. Democratic Technology, Population, and Environmental Change. In *Democratizing Technology: Andrew Feenberg's Critical Theory of Technology*, edited by Tyler J Veak. Albandy: State University of New York Press, 2006.

Sunstein, Cass. *Risk and Reason: Safety, Law, and the Environment*. Cambridge: Cambridge, UP, 2002.

Sunstein, Cass and Richard Thaler. *Nudge: Improving Decisions about Health, Wealth, and Happiness*. New York: Penguin, 2009.

Veak, Tyler J. Ed. *Democratizing Technology: Andrew Feenberg's Critical Theory of Technology*. State University of New York Press, 2006.

Vogel, Steven. *Against Nature: The Concept of Nature in Critical Theory*. Albany: State University of New York Press, 1995.

———. "Habermas and the Ethics of Nature." In *The Ecological Community*, edited by Roger S. Gottlieb. New York: Routledge, 1997.

United Nations Development Program. Human Development Reports. Accessed Jan. 2, 2017 at http://hdr.undp.org/en/content/human-development-index-hdi.

CHAPTER 8

Between Instrumentalism and Determinism: Western Marxism and Feenberg's Critical Theory of Technology

Matthew Greaves

Introduction

Andrew Feenberg's critical theory of technology presents an innovative challenge to contemporary Western Marxism, which is occupying two equally unsustainable positions. These positions are concisely anticipated in the epigraph to Michael Hardt and Antonio Negri's *Empire*, which begins: "every tool is a weapon if you hold it right." The meaning Hardt and Negri intend of the lyric, from Ani DiFranco's song *My IQ*, becomes clearer as *Empire* progresses. Our personal but networked digital technologies, striking in their banality, may be turned against capital or oppressive states if wielded strategically. While the creativity of users is primary in this account, the phrase implies a form of instrumentalism: the socialist

This paper is a variation of one I published in 2015. See Matthew Greaves, "The Rethinking of Technology in Class Struggle," *Rethinking Marxism* 27 no. 2 (2015): 195–211.

M. Greaves (✉)
Simon Fraser University, Burnaby, Canada

D.P. Arnold and A. Michel (eds.), *Critical Theory and the Thought of Andrew Feenberg*, DOI 10.1007/978-3-319-57897-2_8

project can proceed with current digital technologies because what a tool may do is determined by the intentionality of its user.

For the aphorism to be comprehensible in Marxian technology theory, emphasis should indeed be placed on its second half. The location of purposeful activity in users highlights an ability to act meaningfully in technical networks too often foreclosed in theories of technology in class struggle. Yet the inversion raises questions of technical practice that Hardt and Negri do not adequately resolve. It is unclear, for example, how complex instruments that support what Jodi Dean calls "communicative capitalism" may be reordered or recoded to facilitate a power expressed by subordinate groups. It may therefore be objected, as Hardt and Negri expect,[1] that the tools of Empire are those of command and exploitation rather than liberation.

Opposing Marxian conceptions of technology from Dean and Christian Fuchs inhabit this position. Both Dean and Fuchs put forward theories of technology that emphasize command and political foreclosure upon digital networks, which, in what follows, I connect to readings of Marx and Herbert Marcuse. By locating a thread running from Marx to Marcuse then to Dean and Fuchs, my aim is to present a genealogy of what I call foreclosure theory that charts shared ideas. I show that Dean's communicative capitalism hypothesis shares a common logic with the Frankfurt School critique of technological rationality and attendant postulations of working-class political inactivity. The thread of foreclosure theory that connects Marcuse and Dean includes nominally inclusive ideological and social forms that soften the experience of capitalist exploitation. Marcuse believed that the revolutionary potential of the working class was undeveloped, owing to its privileged position within the manufacturing system. Dean, similarly, argues that the hegemony of communicative capitalism deprives struggle of politics because communicative activity on digital networks tempers vital antagonisms. Articulations of class struggle must then occur from without, "as capitalism has subsumed communication such that communication does not provide a critical outside."[2]

The thread linking Marcuse and Fuchs is more easily identified. Through technology, both identify a potential to realize free, nonalienated human activity.[3] But in lines of development organized by capital, technology is encoded for exploitation and social control. The promise of participation that follows digital media is therefore a lie. The individual user is estranged in her activity, laboring for capital under alienated conditions. But *pace* technological substantivism, appropriation and reform of technology is still possible. For Fuchs, the necessary condition for digital technology to achieve its inherent participatory potential is communalization. His construct is unevenly applied, however, as lines of socialist development only emerge from networks collectively designed and run outside of market imperatives. Any potential for alternative rationalities to materialize in technology through articulations of struggle from below is foreclosed upon networks encoded by capital, which defang class struggle.

While both Dean and Fuchs provide welcome critiques of digitally networked communication in contemporary articulations of class struggle, their accounts of technology politics allow very little room, if any, for contestation to influence technological and social development of these networks. Foreclosure theory holds an impaired form of class struggle, with collectives incapable of imposing changes from below on technical networks. While Dean and Fuchs are correct that exploitation and social control occur through digitally networked technology, these same forces create contradictions among a digital proletariat invested with a high degree of technical competency.

Feenberg's critical theory of technology moves in between the poles of instrumentalism and foreclosure theory. The fetishism of technology is uncovered in a materialist form that stresses interaction between abstract social forces and local elements. Feenberg's dialectic captures the relation of ownership, social or private, to local structures in paths of technological development without assuming determination by the former. Following the genealogies of foreclosure theory, I develop through Marx and Feenberg an alternative, yet provisional, theory of digital technology and class struggle. I argue that Hardt and Negri's formulation of the post-Fordist subject suggests intriguing possibilities for theories of class struggle and technological development. Yet any theory of technology based in Hardt and Negri requires, at the very least, tempering by critical theory. Feenberg's work on reification and agency in technical practices presents a particularly compelling corrective. The development of this model presents its own problems, however. Feenberg's critical theory of technology fits awkwardly with the emphasis on actualization through difference that Hardt and Negri draw from Deleuze, requiring a retreat from the multitude concept and Deleuzian schema more generally. The Deleuzian rejection of political inscription is irreconcilable with Feenberg's critical theory of technology, which uncovers the social origins of technological biases. The aggregative Deleuzian system that underpins Hardt and Negri's combinations of the human and the technical spurns the sort of capitalist encoding of technology that Feenberg's concept of democratization[4] is meant to challenge.

Indeed, the critical theory of technology opposes the participatory content of technology against capitalist inscription. The constructivist insight that the immediate social world is reflected in technical systems means that barriers to the socialist transformation of technology are indeed erected. These barriers are, however, surmountable. If the social world is reflected in the technical, as foreclosure and critical theory alike have it, so too is social action, not simply ownership. "Men make their own history, but they do not make it just as they please; they do not make it under circumstances chosen by themselves, but under circumstances directly encountered, given and transmitted from the past. The tradition of all the dead generations weighs like a night-mare on the brain of the living," Marx notes in *The Eighteenth Brumaire of Louis Bonaparte*.[5] Users still hold subordinate positions in technical systems,[6] yet, as in constructivism, "technology remains a site of social struggle," for Feenberg.[7] The past reflected in technology and technical systems, what Feenberg calls technical code, limits action, usually to boundaries acceptable to capital. Technological limitations are permeable, however, and the fundamental inequality at the heart of capitalism produces counter-hegemonic tendencies, especially among a technologically competent proletariat, suggesting new, democratically oriented paths along which technology might progress, which a concluding discussion of Johan Söderberg's work will show.

Given that the processes of entering the world and into relations with individuals render technologies and technical networks political, as Feenberg's work intends to demonstrate, the final section of this paper is meant to upset both Deleuzian schemas and foreclosure politics by conceiving of technology itself as, predominantly, an epiphenomena of class struggle, which Feenberg's work allows. The critical theory of technology discovers agencies embedded in technical networks, each the product of a plurality of social forces. The codetermination of user and tool, positioned by extant constraints yet constitutive of the political horizon, resists both instrumentalism and foreclosure politics. It exists in between the absolute creativity invested in the spontaneous postmodern subject of Empire and theories of a foreclosed political horizon. Although this point may seem banal, position at one or the other pole has characterized enough scholarship to make it relevant, likewise the intervention of Feenberg's work within Western Marxism.

Singular Bodies Seeking Relation

The neutral, instrumental coding of technology in Hardt and Negri's work poses overlapping theoretical and practical problems for the socialist project. Theoretically, the question arises as to whether Hardt and Negri retreat from the class bias of capitalist technology that Marx discovers in factory machinery and which the Frankfurt School, among others, realize had spread to capitalist society at large. Practically, this question hinges on the ability of users to make use of digital technology for socialist change and the degree to which the tools themselves allow a noncapitalist horizon. The point of distinction, here, is the mediating presence of social codes and contingencies materialized in technology and technical networks, and whether, if present at all, these codes are determining or perhaps codetermining of individual action or class struggle.

The questions that arise but remain unresolved because of an implied technological neutrality are fundamentally important given the technological composition of the Empire that Hardt and Negri uncover. Expansive digital networks underwrite Empire's status as the qualitatively dominant form of millennium-era capitalism. The shift from factory production is constituted technologically in the fabric of communication networks. The pace of communication is sped up, enabling ever-tighter feedback loops of information. Nodal points multiply, as do connections between these points, and the subject is oriented by these proliferations, which Lyotard observes.[8] The tools of Empire that support intensified interactivity at the same time facilitate its capture and rendering. Commodifiable forms of communication increase because the connective fabric of informationalized production is organized toward the valorization of communication and cooperation.

Empire subsumes more and more moments of life, as the proliferation of connected nodes facilitates the gathering together of unique situations under its conditions. Appropriately, then, its political project sees radical social change in concrete and singular moments of struggle, immanent to postmodern production and within a collective telos.[9] In *Declaration*, Hardt and Negri identify this relationship with the global outbreak of struggles that began the year prior. Protests across the globe had endemic aims yet were in communication, learning from one another by localizing broader tactics. "The Egyptians, of course, clearly moved down paths traveled by the Tunisians and adopted their slogans, but the occupiers of Puerta del Sol also thought of their struggle as carrying on the experience of those at Tahrir. In turn, the eyes of those in Athens and Tel Aviv were focused on the experiences of Madrid and Cairo. The Wall Street occupiers had all of them in their view."[10]

While antagonisms exist in the field of communication, the supporting technologies configured with larger networks are constitutive all the same of a self-making subject, of the local struggles of the multitude (the exploited social body in Empire), and of the larger telos, per Hardt and Negri. In the global outbreak of struggles, tools—like tactics—are neutrally coded, despite comments indicating otherwise.[11] The materialization of class relations in technology is the opposite of determining[12]; tool users reconfigure networked relations, aggregating preferential nodal points and reconstituting communicative networks toward emancipatory struggle.[13] Globally linked computer technology acts as the passageway through which individuals may actualize new forms of society. The multitude produced by Empire "becomes structure not by negating the originary productive force that animates it but by recognizing it; it becomes language (both scientific language and social language) because it is a multitude of singular and determinate bodies that seek relation."[14]

Technology and the Fantasy of Participation

It's a similar incorporation of political activity on digital networks that Dean believes robs the Left of collective power. Dean's theory of communicative capitalism has Left communication unexpectedly support capitalist exploitation and social control when expressed in digital networks. Operating within the sphere of communicative capitalism, users are routed toward politically empty ideas and projects rather than collective, emancipatory goals. The multitude of Empire, at the opposite pole to Dean, is an included power. "The multitude not only uses machines to produce, but also becomes increasingly machinic itself, as the means of production are increasingly integrated into the minds and bodies of the multitude."[15] Empire is, then, not a system of exclusion "but one of differential inclusion."[16] The aggregative schema that supports Left politics in Empire produces an expansive notion of the exploited class that includes everyone and their tools, as capital discovers ways to commodify most daily interactions. For Dean, this universal inclusion within Empire deprives the multitude of the antagonisms that politics require, resulting in a foreclosure that can only be rectified through external politicization. In Empire, however, there is no outside, no exclusion upon the digital terrain, and therefore no potential external to render the multitude political. In the communicative capitalism hypothesis, it is precisely the point of exclusion for "the alien, the foreigner, the outsider, the rabble, or the multitude," at which disturbances may occur.[17] The immediate conditions and experience of exclusion here recall Marx's political interest in alienation. Under capitalist organization, the estrangement of natural correlates, including that of self-determination, emerges as a condition of existence.[18] Given our alienated conditions of existence, the political activity of the working class is revolutionary, as we struggle to overcome the capitalist system.

When it occurs through the networks of communicative capitalism, emancipatory political activity is deprived of its properly radical character, Dean argues. Identification with false potentials is paradigmatic of communication on digital networks. "Ideals of access, inclusion, discussion, and participation come to be realized in and through expansions, intensifications, and interconnections of global telecommunications. But instead of leading to more equitable distributions of wealth and influence, instead of enabling the emergence of a richer variety in modes of living and practices of freedom, the deluge of screens and spectacles undermines political opportunity and efficacy for most of the world's peoples."[19] Rather than serving as the point "at which a system can be disrupted," digitally networked communication incorporates individuals into nominally democratic forums. This movement acts as part of a double foreclosure. Users are separated from positions of real power and points of resistance are eliminated through a form of inclusion that eliminates challenges to power through cooptation.

The inclusive, mollifying system of communicative capitalism also resembles the critique of postwar discursive structures in Marcuse's technologically rationalized society. Discursive transformations in the era of advanced industrial society promote what Marcuse calls one-dimensionality. In a society organized by technological imperatives, syntax loses its negative character, as the apparent foreclosure of the Marxist revolutionary telos by history is tied to the ideologies of modern, technology-advanced capitalism. In *One-Dimensional Man*, revolution is displaced from working-class activity, in part through the corruption of language. Critical capacities are overtaken by the immediately given; discourse is defined by technologically constituted rationality. Marcuse identifies this transformation with the spread of what he calls the "language of total administration."[20] Postwar social discourse lost contradiction in its language, reflecting a society "increasingly capable of satisfying the needs of the individuals through the way in which it is organized."[21] In a passage that appears to marry Heideggerian technological substantivism and Western Marxism, the reification identified by Georg Lukács takes on the character of postwar industrial technological capitalism. "The organizers and administrators themselves become increasingly dependent on the machinery which they organize and administer. And this mutual dependence is no longer the dialectical relationship between Master and Servant, which has been broken in the struggle for mutual recognition, but rather a vicious circle which encloses both the Master and the Servant."[22] Discourse reflective of technologically advanced production "testifies to identification and unification, to the systematic promotion of positive thinking and doing, to the concerted attack on transcendent, critical notions. In the prevailing modes of speech, the contrast appears between two-dimensional, dialectical modes of thought and technological behavior or social "habits of thought."[23] The superstructure is then productive of a particular working-class subject that moves abreast with technocapitalist imperatives. Communication becomes defined by a crude operationalism, or operational rationality, as the thing and its function are conflated in thought.[24] Commercial and political expression promote "the immediate identification of the particular with the general interest, Business with National Power, prosperity with the annihilation of potential."[25] The negative potential of social actors, realigned positively with the functioning of capital, is now associated with the satisfaction of needs by industry, as technological solutions are accepted for social problems.

For Marcuse, the socialist horizon is foreclosed because the needs of workers and the means for their fulfillment are reconfigured to suit the demands of the capitalist form of technological rationality. The once critical rationality held in the oppositional strata of workers is incorporated into the industrial apparatus, becoming one-dimensional. While the condition of the working class within the industrial apparatus is one of political foreclosure, Marcuse sees potential for a critical rationality elsewhere, though it remains undeveloped. In "Some Social Implications of Modern Technology," originally published in 1941, he writes that "[critical] rationality can fully develop only in social groups whose organization is not patterned on the [industrial] apparatus in its prevailing forms or on its agencies and institutions."[26]

This thought is echoed in *One-Dimensional Man*, where he argues that "the totalitarian tendencies of the one-dimensional society render the traditional ways and means of protest ineffective—perhaps even dangerous because they preserve the illusion of popular sovereignty," whereas "the substratum of the outcasts and outsiders, the exploited and persecuted of other races and other colors, the unemployed and the unemployable" pose revolutionary opposition, if not consciousness.[27] While the administrative tendency remains dominant in one-dimensional society, its containment is not total. Modern society produces politically sublimated workers. However, it likewise produces the countertendency of subversive rationality among those outside the industrial apparatus. "Forces and tendencies exist which may break this containment and explode the society."[28]

Reading backward, the foreclosure of working-class politics in advanced industrial capitalism is threaded through Dean's Left politics on digital communications networks. Marcuse's identification of superstructural elements that condition the needs of workers toward the demands of the industrial base, the historical experience of technological rationality, and the prevailing one-dimensional character of modernity all foreclose political activity, anticipating Dean's communicative capitalism hypothesis. Cooptation is indeed a hallmark of communication on the networks of communicative capitalism. The subsumption of communication in communicative capitalism mollifies the experience of exploitation and alienation, augmenting capitalist and state power at the expense of truly democratic organization. The market, and not collective political activity, then appears as the site of democratic opportunity, presenting users a semblance or fantasy of participation, when the possibility has already been foreclosed.[29] "Communication technologies contribute to the displacement and dispersion of critical energy such that even as inequality has intensified, forming and organizing a coherent opposition has remained a persistent problem—and this in a setting lauded for the way it provides everyday people with new capacities for involvement."[30]

A distinct moment of foreclosure occurs through circularity itself, to the intensity of message circulation. The circulation of content is, for Dean, separate from official politics, and the volume of messages flowing through digital networks fragments politics from below. "Instead of engaged debates, instead of contestations employing common terms, points of reference, or demarcated frontiers, we confront a multiplication of resistances and assertions so extensive that it hinders the formation of strong counter-hegemonies. The proliferation, distribution, acceleration, and intensification of communicative access and opportunity, far from enhancing democratic governance or resistance, results in precisely the opposite, the post-political formation of communicative capitalism."[31] Attempts at political communication in communicative capitalism may gesture toward radical opposition but lack common goals amid the diffusion of ever more parochial forms of individual political expression.

Flows of information and communication are therefore characterized by a dialectical move in which exchange value overtakes use value.[32] Emptied of use value, buried under the weight of endless inputs, our contributions locate us within exploitative circuits of capitalist accumulation without providing communicative use value. "The specific contribution has no symbolic efficiency; rather, it marks only the fact of its having been made. This decline in a capacity to transmit meaning, to symbolize beyond a limited discourse or immediate, local context, characterizes communication's reconfiguration into a primarily economic form. It produces for circulation, not use."[33] The effect is the reproduction and solidification of capitalist social relations through networks that the Left mobilizes to challenge exploitation and social oppression.

In the articulation of class struggle, communicative capitalism is above all a form of suppression through cooptation. Digital forums exhaust radical energy in politically defanged projects, rendering it passive. As with Marcuse, containment isn't total. The outside(r) remains. For example, the intervention of Occupy, the subsequent political landscape, and the collective determinations out of which subjectivities form from common terms did not express continuity or homology with communicative capitalism. Instead, these processes disrupted the flow of the Left through networked exploitation.[34] Unlike the multitude of Empire, the properly Left oppositional strata are here formed through the articulation of collective terms and goals outside of the digital networks that constitute communicative capitalism.

Technological Change: Relations of Production or Class Struggle?

Like Dean, Fuchs believes that the promise of participation that follows new media is false. Users instead act like laborers, producing value through networked communication and cooperation with fellow users. New media does not help constitute revolutionary subjectivity but transforms user inputs into value, as we perform a kind of unwaged labor for the owners and investors of new-media sites. Drawing from Marcuse, a free and creative relation to technology remains available. Digital communications technology is said to be ambivalent. Yet the shaping of platforms by capitalist forces produces minimal opportunities for genuine communication. While a correct assessment of exploitation, Fuchs overplays his hand, and, in doing so, the theory of technical foreclosure on digital networks he develops lapses into a determinism that the gesture of ambivalence means to avoid.

Revisiting Dallas Smythe's audience commodity thesis,[35] Fuchs believes that new-media users are actively produced through digital networks and sold to advertisers in a process that commodifies their activity. The lack of common ownership or user input in decision making means that the communication between users and the production of content do not reach the threshold for participation. The primary relationship is instead one of exploitation, domination, and commodification, in which communication is shaped toward the production of value. Any political information or discussion that occurs is overwhelmed by this shaping, which privileges entertainment in the production of audiences. For Dean, politics in the era of communicative capitalism may be retrieved through external organization. Fuchs, on the other hand, identifies the possibility of a "participatory, co-operative and sustainable information society (PCSIS)"—formulated more concisely as a communist Internet—based upon communal values and ownership. "A democratic communication infrastructure requires strengthening the communication commons. The task is to advance communist media and a communist Internet in a democratic and participatory communist society."[36]

Fuchs's emphasis on social ownership as a condition for participation is further developed in his revised base-superstructure schema. Positing a departure from direct subordination by the base in the work of Marxist cultural theorists (he cites Raymond Williams and Stuart Hall), the superstructure is presented as "a complex, nonlinear, creative reflection of the base, the base a complex nonlinear creative reflection of the superstructure."[37] Unlike the formulations of cultural theorists, Fuchs asserts that there is nothing mechanistic about his base-superstructure relation. The base is composed through exchanges between nature, technology, and labor mobilized toward the satisfaction of human need. The superstructure exists at the junction of political and cultural systems, through which "'Immaterial' goods emerge, which allow the definition of collective decisions and societal value structures."[38]

Further complicating the once simple metaphor, base and superstructure appear *horizontally* related through a *trichotomy* of economic, political, and cultural subsystems, each communicating and affecting the others through feedback loops.[39]

According to Fuchs, a symmetrical distribution of participation acts as the threshold for a participatory social order supported by technological subsystems. Given the relative autonomy of base from superstructure, implicating the superstructure in participatory regimes requires that Fuchs develop an equally nonmechanistic means of transformation, and he finds such a model in Nancy Fraser and Axel Honneth's dialectic of recognition and redistribution. Fuchs argues that recognition is the redistribution of symbolic power. It is the superstructural form of economic redistribution, a process of cultural recognition and communication. Redistribution within the base is economic transformation. A lack of recognition in the superstructure means inequitable representation conforming to existing social asymmetries and attendant norms. Crucially, economic redistribution implies more just recognition following naturally from equitably reconfigured economic relations. Redistribution is incomplete without the abolition of social hierarchies, and in this way cultural recognition or redistribution is meaningless without the abolition of class. To be complete, recognition in the superstructure requires redistribution in the base.

The relative autonomy of the superstructure from the base appears to save Fuchs from determinism, in this instance. The content of the extant superstructure—the symbolic order of capitalism—does not mechanically act to sublimate revolutionary desire. Yet behind the symbolic order there still lies the linchpin base. Participation, which requires recognition, cannot occur without redistribution in the base. And the superstructure appears, as it does classically, to legitimize economic and political systems within a recursively linked though indeterminate structure. While ambivalence is acknowledged, an ossified productive relation appears again as the determining principle of technical networks and digital communication, precluding participatory communication from taking place.

This schema appears throughout Fuchs's work. Technological ambivalence, for example, resides in an approach of mutual shaping between technology and society. "A critical theory of technology is a specific mutual shaping approach that adds the idea that technological development interacts with societal contradictions."[40] Antagonism, exploitation and contradiction are thus said to materialize in ambivalent technology and technological processes. Contradictions on social media appear between the individual opportunities afforded by social networks and their function as advanced surveillance apparatuses.[41] Yet the classic determination remains intact. Digital networks are riven with contradictions that are resolved or mollified in acceptable ways. Radical activity is politically muted on networks encoded by capital, as the symbolic, discursive order reflected the needs of capital.

Although in a more complex form, Fuchs and Sevignani's *becoming-art* of the economy and technology, based on the work of Marcuse and William Morris, also repeats this schema. Fuchs and Sevignani believe that, when directed away from material production and toward human necessity, technology and mechanization may be redefined as the elimination of want. In a cooperative and unalienated society, technology—especially laborsaving technologies—might eradicate necessary but toilsome labor. Communist organization could develop laborsaving technologies that would promote "creative work and the becoming-art of the economy."[42] Through the abolition of toil, work would come to reflect the creative capacities of all individuals. As Marcuse similarly writes: "Utopian possibilities are inherent in the technical and technological forces of advanced capitalism and socialism: the rational utilization of these forces on a global scale would terminate poverty and scarcity within a very foreseeable future."[43] To this Marcuse attaches a caveat that Fuchs repeats: "But this *gaya scienza* is conceivable only after the historical break in the continuum of domination—as expressive of the needs of a new type of man."[44]

The form of rupture through which we might see technology develop toward the satisfaction and realization of human qualities requires what Marcuse calls an "ingression of the future into the present,"[45] which he identifies with anarchistic and rebellious movements. To be sure, Marcuse has in mind specific tendencies within liberation movements of the 1960s, but the emphasis on spontaneity suggests that radical activity may act upon exploitative technologies and networks. To read this ingression at the exclusive level of broadscale revolutionary change ironically limits its transformative function.

For Fuchs, technology built upon the foundational class distinction of capitalist production is ambivalent, situated between multiple paths of development and co-constitutive or mutually shaped by dialectical processes.[46] However, mutual shaping may only incorporate proletarian activity within extant communist or commonist architecture. Otherwise, a static productive relation determines the technical. Noncommercial social media sites such as N-1 and Occupii, by contrast, support the growth of the communist/commonist Internet. Their architecture suggests a flourishing non-commodified commons, with communication and cooperation structured toward collective values and common struggles. N-1 summarizes its political relevance through Audre Lorde's famous aphorism: "The master's tools will never dismantle the master's house."[47] The popular tools of digital communication cannot be those of capitalism's gravedigger, according to Fuchs and N-1. Digital networks and the content therein support and reflect capitalist orthodoxies, and participation on these networks requires the absence of class distinctions manifest in technology. The foreclosure of politics in Fuchs therefore dismisses technological change in noncommunist media that occurs prior to economic redistribution, while Dean positions communicative capitalism as a politically inoperative zone for Left politics. Fuchs in effect holds that technology requires release from capitalist relations of production to develop in accord with human potentialities. The Communist Internet has communalized property relations subordinate nominally ambivalent technology as it develops. While a generally compelling argument, that proletarian struggles on digital networks materialize in lines of technological development is absent from this position.

Critical Technological Change

The phenomenology of the foreclosurist position, in which the subject, like their tools, are constructed in the image of capital's needs, is in direct opposition to the Deleuzian ontology of Empire. Although it's a reduction, this distinction may be instructively characterized as moments of mediation and negation in critical theory and aggregative multiplicity in the empiricism of Deleuze. Fuchs and Dean allow room for the exploited to negate the conditions of their exploitation, although neither sees much of it on the digital networks that support postmodern accumulation. At the opposite pole, Hardt and Negri believe that the technology of postmodern capitalism helps constitute the multitude itself, and they identify a possibility for this technology to act as a passageway for radical desires. More than this, the division between humanity and technology evanesces in the political project of Empire, as they culminate in a politically oriented hybrid unit.

> The fields upon which productive forces are reappropriated by the multitude is a field of radical metamorphoses—the scene of a demiurgic operation. This consists above all in a complete revision of the production of cooperative subjectivity; it consists in an act, that is, of merging and hybridizing with the machines that the multitude has reappropriated and reinvented; it consists, therefore, in an exodus that is not only spatial but also mechanical in the sense that the subject is transformed into (and finds the cooperation that constitutes it multiplied in) the machine.[48]

The multitude can become a revolutionary body because of this hybrid constitution. Its cooperative potential, beginning with singular appropriations of technology, spreads because of a multiplication of hybridized nodal points. As these points swell, the multitude recognizes in its reappropriations the false conflict constructed between humanity and technology, as Marx describes in *Capital.* "It took both time and experience before the workers learnt to distinguish between machinery and its employment by capital, and therefore to transfer their attacks from the material instruments of production to the form of society which utilizes these instruments."[49]

The evanescent human–nonhuman relation Hardt and Negri develop is perhaps equal parts Marx and Deleuze, although not the Marx of *Capital*'s fifteenth chapter, despite the quote, nor is it necessarily consistent with autonomist Marxist thinking on the factory, which forms the basis of Hardt and Negri's position. As Matteo Pasquinelli has shown, the Italian autonomist Romano Alquati identified the materialization of class biases in the cybernetic factory as early as 1962, in his study of Olivetti.[50] Like the digital networks that characterize the immaterial labor hypothesis, Alquati discovered that workers' cognitive and creative acts materialized not only in products but also in machinery and labor processes. "The proletarian here," as Pasquinelli puts it, "is no longer just a thermodynamic animal steaming in front of a machine but is already a brain worker."[51] But Alquati concluded from his research at Olivetti that capitalist despotism "realizes itself above all through its technology, its 'science,' the diffusion of its structures of class exploitation in social life, through constant capital which embraces all," in and out of the factory.[52] Similar to Hardt and Negri's *Empire*, Alquati identifies the inputs of worker information in the technical functioning of the cybernetic factory.[53] Yet the status of postmodern technology and the proletarian–technical relation is generally less antagonistic for Hardt and Negri—who deploy Marx's observations on the Luddites and other machine breakers, cited above[54]—than it is for Alquati in the cybernetic factory. The mediation of machinery at Olivetti appears different than the effectively neutral conception of networked digital technology.

Diffused from the factory, the total generalization of technology across the social field suggests multiple determinants not immediately given to nor *of* the multitude. We might here recall a commonly quoted observation from *Capital*: "It would be possible to write a whole history of the inventions made since 1830 for the sole purpose of providing capital with weapons against working-class revolt."[55] By contrast, autonomist theorists like Pasquinelli have progressed toward a "liquefaction" of the machine, reading contemporary capitalism as its transposition in the body and in the intellect of labor.[56] A favorite of the Italian autonomists, Marx's economic notebooks of 1857–1858, gathered together as the *Grundrisse* draft, may to a degree suggest aspects of these theoretical positions. The growth of a vast technical apparatus encourages the release of creative capacities in the social body, the "artistic, scientific etc. development of the individuals."[57] The reduction of human labor to a minimum posits labor power as a poor source of value for capital, displacing the labor theory of value in its classic Marxist formulation. Instead, general social knowledge is constituted as the dominant basis of wealth, instead of labor power. Some of this appears to correspond to the valorization of communication and the attendant evaporation of socially necessary labor time that arguably characterize the accumulation of surplus value through user activity on digital networks. To engender radical potential, however, and to move beyond "the material conditions to [simply] blow" the foundational conditions of capital "sky-high,"[58] Hardt and Negri require a further step. Repudiating dialectic movement in their framework of positive becoming, they attempt to integrate Marxian categories into an aggregative Deleuzian logic. The introduction of Deleuze is a rejection of the dialectical formula that typically directs Marxist thought.[59] It is meant to guard against negation or the overcoming of exploitation through history. The positive method of Deleuzian becoming is one of distinction and coalition, of becoming something else through difference and aggregation.

In *Declaration*, struggle travels and mutates as it cycles through different locations. Revolutionaries in Egypt and Tunisia, occupiers in North America and Europe all reject political representation in favor of multitudinous, leaderless organization. Democratic principles are achieved through collective determination and horizontal decision making. Groups do not evolve by negation or representation but rather "spread out like swarms," actualizing in positive acts.[60] Struggle, rather than taking on a dialectical character, emerges and proceeds through coalition building, as local struggles are articulated within a common framework and communicate with one another through communicative technology. Agency appears here in connected, hybridized subjects and their entanglements. Gone is the working class, bearing the weight of universal liberation from capitalism. In its place is something like permanent revolution dictated by a multiplicity of singular yet connected bodies that struggle toward irreducible ends.

The difficulty with Hardt and Negri's use of Deleuzian aggregation, from the perspective of Marxian theories of technology, is twofold. First, the condition of technological neutrality under which the multitude takes hybrid form displaces critical attention that should be paid to technology. The multitude, we'll remember, proceeds not through negation of potentially antagonistic or exploitative technology. Rather, its successes come through the recognition of the radical potential found in these human–technical hybrids[61]: The multitude incorporates a "set of powers to act (being, loving, transforming, creating)."[62] The inseparable co-constitution, however, robs digital technology of the critical attention it deserves. Second, the multiplication of nodal points is an endlessly destabilizing force. Perpetual flux deemphasizes the social qualities that influence technological development and, as Dean argues, results in fragmentation of the formerly collective determinants of social change. Reviving a critical account of technology among the Left thus requires deviation from Hardt and Negri's hypothesis.

At this point the philosophy of technology becomes relevant, as the question of how to critically account for technology in social struggle without lapsing into instrumentalism moves to the forefront. Feenberg's work on agency and integration in socio-technological fields (fields that tend toward exploitation and social control) is a critical perspective that may dovetail the technologically competent radical of Hardt and Negri. But to put Feenberg in communication with Hardt and Negri requires the multitude concept present as at least momentarily stable. Without this, criticality loses coherence amid ever more important connections.

Feenberg emphasizes the combinations that technologies and subjects form, combinations that exceed the immediate user–technology relation. Technical action in contemporary capitalism requires integration with multiple imbricated systems—social, technological, vocational, political, and natural, to name a few. Pretensions of user autonomy are thus confirmed as such by engagement with the technical object at immediate and systemic levels. In its most basic form, the user–technology relation is a codetermination, a mutual conditioning of subject and object. Expanding outward, further determinates begin to appear. Corresponding norms and dispositions code individual activity within overlapping frameworks.[63] The individual user is produced by way of multiple intersecting relations as she meets the conditions for her enrollment. With digitally networked communication, the local customs of different communities suggest potential for Left spaces. Yet processes of subject integration, in positioning the subject as community member, do not necessarily overwhelm the reifying moments objects pass through before realization. To be successfully ordered in a technical system, objects in transformation are subordinated to the functioning of existing systems and "reduced to those aspects through which they can be enrolled in a technical network."[64] At the same time, as the object is being manipulated to fit within technical networks, its reduction to a set of workable qualities implicates socially constructed biases in the process, enhancing certain possibilities inherent in the object while precluding others.

For Feenberg, the social and the sociotechnical are multiple and contested. The technology–user relation is not defined by an immediate nor ultimate dualism; multiple social constructions are irreducibly present from the beginning of struggle, as human and nonhuman come into relation within politically relevant networks. This understanding corresponds, moreover, to a plurality of social situations. "Struggles emerge around many issues, all of them traversed by technical mediations, but only a few of them primarily labor issues"[65] (at least *waged* labor, Feenberg might have added). If it is to trouble technological domination, resistance must take seriously multiple embedded forms of exploitation and alienation. "Only a democratically constituted alliance of actors, embracing [a multitude of affected social groups], is sufficiently exposed to the consequences of its own actions to resist harmful projects and designs at the outset."[66]

In *Transforming Technology*, Feenberg produces a dialectic of abstract and concrete moments, realized in the lines of development that produce technological change, in which there is a definite relationship between abstract capitalist tendencies and local phenomena. Here, Feenberg theorizes potential investments of solidaristic and communal values within technical networks.[67] That capitalist history, codified in the technical, weighs upon the minds of the living suggests digital networks are prone to exploitative and reactionary orientation. Indeed, that technical networks embody socially constructed biases demonstrates as much, developed as they are through capitalist social relations. Drawing from both Marx and the actor-network tradition, Feenberg suggests that control of users by capital is embedded in technical networks through an imbricated process of instrumentalization for both user and technology.[68] To be successfully configured in a network, objects are positioned and assigned meaning through socially determined value systems embedded in networks, both functionally and ideologically.

In this way, technical networks exert forms of control over users. The inverse is, however, also true. Technologies allow user manipulation in a variety of ways. This relationship is the core of Feenberg's dialectic. In use, subject and object form an entangled hierarchy, through which identity and meaning are challenged and solidify.[69] Control in the immediate relationship of subject to object coheres in a two-level process. Abstract economic tendencies begin to define the relationship between users and technology. The concrete entangled hierarchy is the immediate relationship between user and technology, delimited though not determined by capital. This ambivalence of user and technology suggests that technology and technical networks may be recoded to embody socialist values, while admitting that technical networks may also push users toward exploitative or reactionary experiences. The relation of abstract to concrete allows some degree of maneuverability for those exposed to a given technical network. The exact terms in which technical networks may be engaged are however determined by history and locality. And the concrete reality of networked interaction is itself structured by the multiple reductive processes that user and technology undergo. In an 1890 letter to Joseph Bloch, Friedrich Engels clarifies the relationship of abstract to concrete in historical materialism, in a manner that is also instructive for the critical theory of technology.

> According to the materialist view of history, the determining factor in history is, *in the final analysis*, the production and reproduction of actual life. More than that was never maintained either by Marx or myself. Now if someone distorts this by declaring the economic moment to be the *only* determining factor, he changes that proposition into a meaningless,

> abstract, ridiculous piece of jargon. The economic situation is the basis, but the various factors of the superstructure — political forms of the class struggle and its consequences, namely constitutions set up by the ruling class after a victorious battle, etc., forms of law and, the reflections of all these real struggles in the minds of the participants, i.e. political, philosophical and legal theories, religious views and the expansion of the same into dogmatic systems — all these factors also have a bearing on the course of the historical struggles of which, in many cases, they largely determine the *form*.[70]

The constitution of user and technology by the economic moment and concrete historical forms of the superstructure has as a correlate the creative enactment of tactics against control, as Marcuse (20, 89) suggests. Power is enacted and shifted as groups attempt to favorably reconfigure technical fields. But creativity does not dispatch with the reification of technology and user. The reduction to primary forms limits behavior and use, allowing for some possibilities and not others. Such stability, while denying absolute autonomy and creativity, suggests possibilities for socialist struggle.

Söderberg's figure of the technological misuser corresponds to the Feenberg/Marxian model of technical change, in between instrumentalism and determinism.[71] The misuser identifies a broadly constituted political subject produced through a series of antagonistic relations that similarly condition technological development. The intervention of hackers in software code, for example, suggests not only potential subversions of identifiably capitalist codification but also alternative and perhaps subversively communist paths for ambivalent technology to travel.[72] To repeat an earlier criticism, while Fuchs and Sevignani do well to identify nonalienating social networks, the determining economic moment precludes struggle from conditioning technology not already communist/commonist in ownership, what Feenberg calls subversive rationalization.[73] Alternatively, Söderberg's research shows that oppositional moments of class struggle push forward technological development and subjectivity alike.[74] Hackers and other misusers attack alienated activity from within, what we can identify as disalienating, rather than nonalienated, activity. While the lay expertise of hackers perhaps represents something close to that of Hardt and Negri's creative technical subject, the "showcase of play struggle" within the larger rubric of the misuser is based upon direct antagonisms and alienation via mechanisms of political and social control.[75] The actions of misusers occur within new-media platforms said to foreclose political struggle, yet their oppositional activity produces subversive lines of technological development, suggesting that the subsumption of digital communication by capital remains unfinished, a basic point of Feenberg's recent writing.[76]

Conclusion

The critical theory of technology proposes a Marxian method for technological change that moves in between instrumentalism and technological determinism, in which subversion and class struggle may produce disalienating affects. Hardt and Negri's technology politics, by contrast, is rooted in an all-encompassing definition of the multitude, which includes almost everyone and, we can add, their tools. Recognizing multiple points of exploitation and oppression to be overcome means that struggle requires neither the rejection of hitherto capitalist technology nor its passive acceptance. Alternative forms of social organization will of course require us to dispose of some technologies altogether too cumbersome or dangerous. Multiple points of struggle, in short, need to be critical as well. Although digital technology need not become determining, inscriptions of capital in technology require attention.

While providing valuable critiques of digital communication in contemporary capitalism, both Dean and Fuchs overstate the determinacy of the technical in producing subjectivity and alienated activity, as the introduction of Söderberg's misuser means to demonstrate. Digital tools may act as apparatuses of value capture, yet mystification or capitalistic tendencies need not overwhelm or determine technical action. The powers to act inherent in techno-social fields may take place on terrain built by capital, but the character of activity is by no means fixed. Political activities on digital networks may indeed condition technological change, as the misuser identified in Söderberg's work recodifies technological paths through the struggle from below.

For Hardt and Negri, the price of moving beyond determinism is the reduction of digital technology to user intentionality. Their instrumentalism does have the virtue of highlighting the technical agencies of those not holding power, but the position excludes critical analysis of communication technology. They instead conceive of the multitude through a foundational hybrid between technology and humanity. Through distinction between individuals and technology, Feenberg's work suggests a critical form of technological change that allows proletarians to recodify ambivalent technology. While this has only been sketched in rough terms, the path in between determinism and instrumentalism has the virtue of providing a basis from which considerations of activity or struggle in technical fields may better account for class struggle in Marxian conceptions of technology.

Notes

1. Michael Hardt and Antonio Negri, *Empire* (Cambridge, Mass: Harvard University Press, 2000), 359.
2. Jodi Dean, *The Communist Horizon*. London: Verso (2012), 128.
3. See, for example, Christian Fuchs and Sebastian Sevignani, "What is Digital Labour? What is Digital Work? What's their Difference? And why do these Questions Matter for Understanding Social Media?" *Triple C: Communication, Capitalism & Critique* 11 no. 2 (2013), 287 and Herbert Marcuse, "Some of Social Implications of Modern Technology," in *The Essential Frankfurt School Reader*, ed. Andrew Arato and Eike Gebhardt (New York: Continuum International, 2005), 161.
4. Andrew Feenberg, *Transforming Technology: A Critical Theory Revisited* (Oxford: University Press, 2002), 153–155.
5. Karl Marx, *The Eighteenth Brumaire of Louis Bonaparte* (New York: International Publishers), 15.
6. Andrew Feenberg, *Questioning Technology* (London: Routledge, 1999), xiv.
7. Ibid., 83.
8. Jean-François Lyotard, *The Postmodern Condition: A Report on Knowledge*, Jeff Bennington and Brian Massumi, trans. (Minneapolis: University of Minnesota Press, 1984), 15.
9. Hardt and Negri, *Empire*, 403–413.
10. Michael Hardt and Antonio Negri, *Declaration* (Allen, Tex.: Argo-Navis, 2012), 4.
11. Hardt and Negri, *Empire*, 405.
12. Ibid., 367.
13. Hardt and Negri's conception of technology is not one of technologies as such, though digital networks are open to appropriation within polycentric forms of struggle. Hardt and Negri's discussion of military technology in *Multitude: War and Democracy in the Age of Empire*, for example, makes clear that the capacity to reappropriate the digital tools of production is not extended into the realm of war.
14. Hardt and Negri, *Empire*, 30.
15. Ibid., 406.

16. Hardt and Negri, *Multitude*, 134.
17. Jodi Dean, "The Networked Empire: Communicative Capitalism and the Hope for Politics," in *Empire's New Clothes: Reading Hardt and Negri's* "Empire," ed. Paul A. Passavant and Jodi Dean (London: Routledge, 2004), 284.
18. Karl Marx, "Economic and Philosophical Manuscripts," in *Karl Marx: Early Writings*, Rodney Livingstone and Gregor Benton, trans (Harmondsworth: Penguin, 1992) 322-334.
19. Jodi Dean, "Communicative Capitalism and the Foreclosure of Politics," *Cultural Politics* 1 (2005), 55
20. Herbert Marcuse, *One-Dimensional Man: Studies in the Ideology of Advanced Industrial Society* (Boston: Beacon Press, 1991), 85.
21. Marcuse, *One-Dimensional Man*, 1.
22. Ibid., 33.
23. Ibid., 85.
24. Ibid., 87.
25. Ibid., 90.
26. Marcuse, "Some Social Implications of Modern Technology," 149.
27. Ibid., 256.
28. Ibid., xlvii.
29. Dean, "Communicative Capitalism and the Foreclosure of Politics," 55.
30. Dean, *The Communist Horizon*, 126.
31. Dean, "Communicative Capitalism and the Foreclosure of Politics," 53.
32. Ibid., 58.
33. Dean, *The Communist Horizon*, 127.
34. Ibid., 214–6.
35. Dallas Smythe, "Communication: The Blindspot of Western Marxism," *Canadian Journal of Political and Social Theory* 1 no. 3 (1977): 1–27.
36. Christian Fuchs, *Foundations of Critical Media and Information Studies* (Abingdon, UK: Routledge, 2011), 51 and Christian Fuchs, "Class and Exploitation on the Internet," in *Digital labor: The Internet as Playground and Factory*, ed. Trebor Scholz (New York: Routledge, 2013), 213, 221.
37. Fuchs, *Foundations of Critical Media and Information Studies*, 50.
38. Ibid., 50.
39. Ibid., 49.
40. Ibid., 114.
41. Ibid., 120.
42. Fuchs and Sevignani, "What is Digital Labour?" 282.
43. Herbert Marcuse, *An Essay on Liberation* (Boston: Beacon Press, 1969), 4.
44. Ibid., 19.
45. Ibid., 89.

46. Christian Fuchs and Nick Dyer-Witheford, 2013. “Karl Marx @ Internet Studies.” *New Media & Society* 15 no. 5 (2013), 786.
47. N-1 cited in Fuchs and Sevignani “What is Digital Labour?” 271.
48. Hardt and Negri, *Empire*, 366–367; emphasis added.
49. Karl Marx, *Capital Volume One*, Ben Fowkes, trans. (London: Penguin, 1990), 554–555 quoted in *Empire*, 367.
50. Matteo Pasquinelli, “To Anticipate and Accelerate: Italian Operaismo and Reading Marx’s Notion of the Organic Composition of Capital,” *Rethinking Marxism* 26 no. 2 (2014), 182–184.
51. Pasquinelli, “To Anticipate and Accelerate,” 183.
52. Romano Alquati, “Organic Composition of Capital and Labor-Power at Olivetti,” Steve Wright, trans, *Viewpoint Magazine* (27 September 2013), viewpointmag.com/2013/09/27/organic-composition-of-capital-and-labor-power-at-olivetti-1961.
53. Pasquinelli, “To Anticipate and Accelerate,” 183.
54. *Capital Volume One*, 554–555.
55. *Capital Volume One*, 563.
56. Pasquinelli, “To Anticipate and Accelerate,” 188.
57. Karl Marx, *Grundrisse: Foundations of the Critique of Political Economy*, Martin Nicolaus, trans. (London: Penguin), 706.
58. Ibid., 706.
59. “Orthodox Marxism, therefore, does not imply the uncritical acceptance of the results of Marx’s investigations,” George Lukács writes. “On the contrary, orthodoxy refers exclusively to *method*. It is the scientific conviction that dialectical materialism is the road to truth and that its methods can be developed, expanded and deepened only along the lines laid down by its founders.” George Lukács, “What is Orthodox Marxism?” in *History and Class Consciousness: Studies in Marxist Dialectics*, Rodney Livingstone, trans. (Cambridge, MA: The MIT Press, 1971), 1.
60. Hardt and Negri, *Declaration*, 5.
61. Hardt and Negri, *Empire*, 30.
62. Ibid., 357.
63. Feenberg, *Questioning Technology*, 206.
64. Ibid., 203.
65. Feenberg, *Transforming Technology*, 61.
66. Andrew Feenberg, “Replies to Critics: An Autobiographical Note,” in *Democratizing Technology: Andrew Feenberg’s Critical Theory of Technology* (Albany: SUNY Press. 2006), 183.
67. Feenberg, *Transforming Technology*, 131–161.
68. Feenberg, *Questioning Technology*, 205–208.
69. Ibid., 205–207.

70. Friedrich Engels, "Engels to Joseph Bloch in Königsberg, 21–22 December 1890," in *Marx and Engels Collected Work Volume 49: Letters 1890–1892*, (Lawrence and Wishart, 2010), 34–35; emphasis in the original.
71. Johan Söderberg, "Misuser Inventions and the Invention of the Misuser: Hackers, Crackers and Filesharers," *Science as Culture* 19 no. 2 (2010): 151–179.
72. Ibid., 162.
73. Andrew Feenberg, "Subversive Rationalization: Technology, Power, and Democracy," in *Technology and the Politics of Knowledge* (Bloomington: Indiana University, 1995), 3–22.
74. Söderberg, "Misuser Inventions," 154.
75. Johan Söderberg, *Hacking Capitalism: The Free and Open Source Software Movement* (New York: Routledge, 2008), 10.
76. Andrew Feenberg, "Introduction: Toward a Critical Theory of the Internet," in *(Re)Inventing the Internet: Critical Case Studies*, eds. Andrew Feenberg and Norm Friesen (Rotterdam, The Netherlands: Sense Publishers, 2012), 3–17.

Bibliography

Alquati, Romano. "Organic Composition of Capital and Labor-Power at Olivetti." Translated by Steve Wright. *Viewpoint Magazine*. Published September 27, 2013. https://viewpointmag.com/2013/09/27/organic-composition-of-capital-and-labor-power-at-olivetti-1961/.

Dean, Jodi. "The Networked Empire: Communicative Capitalism and the Hope for Politics." In *Empire's New Clothes: Reading Hardt and Negri's* "Empire," edited by Paul A. Passavant and Jodi Dean, 267–90. London: Routledge, 2004.

Dean, Jodi. "Communicative Capitalism and the Foreclosure of Politics." *Cultural Politics* 1 no. 1 (2005): 51–74.

———. *The Communist Horizon*. London: Verso, 2012.

Engels, Friedrich. 2010. "Engels to Joseph Bloch in Königsberg, 21–22 December 1890." In *Marx and Engels Collected Work Volume 49: Letters 1890–1892*, 33–37. Lawrence and Wishart.

Feenberg, Andrew. "Subversive Rationalization: Technology, Power, and Democracy." In *Technology and the Politics of Knowledge*, edited by Andrew Feenberg and Alastair Hannay. Bloomington: Indiana University Press, 1995.

———. *Questioning Technology*. London: Routledge, 1999.

———. *Transforming Technology: A Critical Theory Revisited*. Oxford: Oxford University Press, 2002.

———. "Replies to Critics: An Autobiographical Note." In *Democratizing Technology: Andrew Feenberg's Critical Theory of Technology*, edited Tyler Veak, 175–210. Albany, NY: SUNY Press, 2006.

———. "Introduction: Toward a Critical Theory of the Internet." In *(Re) Inventing the Internet: Critical Case Studies*, edited by Andrew Feenberg and Norm Friesen, 3–17. Rotterdam, The Netherlands: Sense Publishers, 2012.

Fuchs, Christian. *Foundations of Critical Media and Information Studies.* Abingdon, UK: Routledge, 2011.

———. "Class and Exploitation on the Internet." In *Digital labor: The Internet as Playground and Factory*, edited by Trebor Scholz, 211–24. New York: Routledge, 2013.

Fuchs, Christian and Nick Dyer-Witheford. "Karl Marx @ Internet studies." *New Media & Society* 15 no. 5 (2013): 782–796.

Fuchs, Christian and Sebastian Sevignani. "What is Digital Labour? What is Digital Work? What's their Difference? And why do these Questions Matter for Understanding Social Media?" *Triple C: Communication, Capitalism & Critique* 11 no. (2013): 237–293.

Greaves, Matthew. "The Rethinking of Technology in Class Struggle: Communicative Affirmation and Foreclosure Politics." *Rethinking Marxism* 27 no. 2 (2015): 195–211.

Hardt, Michael and Antonio Negri. *Empire.* Cambridge, MA: Harvard University Press, 2000.

———. *Multitude: War and Democracy in the Age of Empire.* New York: Penguin, 2004.

———. *Declaration.* Allen, Tex.: Argo-Navis, 2012.

Lyotard, Jean-François. *The Postmodern Condition: A Report on Knowledge.* Translated by Geoff Bennington and Brian Massumi. Manchester, U.K: Manchester University Press, 1984.

Marcuse, Herbert. *An Essay on Liberation.* Boston: Beacon Press, 1969.

———. *One-Dimensional Man: Studies in the Ideology of Advanced Industrial Society.* Boston: Beacon Press, 1991.

———. "Some Social Implications of Modern Technology." In *The Essential Frankfurt School Reader*, edited by Andrew Arato and Eike Gebhardt, 138–63. New York: Continuum International, 2005.

Marx, Karl. *Capital Volume One.* Translated by Ben Fowkes. London: Penguin, 1990.

———. "Economic and Philosophical Manuscripts." In *Karl Marx: Early Writings.* Translated by Rodney Livingstone and Gregor Benton, 279–400. Harmondsworth: Penguin, 1992.

———. *The Eighteenth Brumaire of Louis Bonaparte.* New York: International Publishers, 1963.

Marx, Karl. *Grundrisse: Foundations of the Critique of Political Economy.* Translated by Martin Nicolaus. London: Penguin, 1993.

Pasquinelli, Matteo. "To Anticipate and Accelerate: Italian Operaismo and Reading Marx's Notion of the Organic Composition of Capital." *Rethinking Marxism* 26 no. 2 (2014): 178–192.

Smythe, Dallas. "Communication: The Blindspot of Western Marxism." *Canadian Journal of Political and Social Theory* 1 no. 3 (1977): 1–27.

Söderberg, Johan. *Hacking Capitalism: The Free and Open Source Software Movement.* New York: Routledge, 2008.

———. "Misuser Inventions and the Invention of the Misuser: Hackers, Crackers and Filesharers." *Science as Culture* 19 no. 2 (2010): 151–179.

PART III

From Heidegger Through Postmodernism: The Ongoing Role of the Critical Theory of Technology

CHAPTER 9

The Question Concerning a Vital Technology: Heideggerian Influences on the Philosophy of Andrew Feenberg

Dana S. Belu

The philosophy of technology is the foundation of all Western philosophy.
Andrew Feenberg

American philosopher Andrew Feenberg was instrumental in bringing about the empirical turn in the philosophy of technology and he did so with the help of the Hegelian–Marxist dialectic. In view of this contribution it might seem odd to connect his thinking with that of Heidegger, whose philosophy, as is well known, consistently bracketed empirical concerns and fiercely rejected dialectical thinking. In this article I will try to show the ways in which aspects of Heidegger's phenomenological ontology, early and late, have influenced Feenberg's social phenomenology of technology.

D.S. Belu (✉)
Philosophy Department, California State University, Carson, CA, USA

D.P. Arnold and A. Michel (eds.), *Critical Theory and the Thought of Andrew Feenberg*, DOI 10.1007/978-3-319-57897-2_9

Inspired by Heidegger's critique of the neutrality of technology yet not persuaded by the asocial ontology of this critique, Feenberg develops a critical theory of technology that attempts to rethink the socio-historical constitution of meaning in accordance with the democratization of technology, i.e., the inclusion of participant interests and the promotion of life-affirming technical design. His work highlights the urgent need for a social and political transformation of technology in accordance with life-affirming values and possibilities. In *Between Reason and Experience* (2010) Feenberg warns that "Once the lessons of experience no longer shape technical advance, it is guided exclusively by the pursuit of wealth and power.... The genocidal twentieth century is now followed by a new century of environmental crisis."[1]

Heidegger's early writings on the phenomenology of lived experience and his later call for a post-technological world, where our relationship to technological use becomes free, shape Feenberg's vision of a phenomenological rationality, a "rational critique of rationality"[2] that is technologically attuned. Feenberg interprets Heidegger's early phenomenology as a phenomenology of action inspired by the model of craftwork and technique. However, he critiques the later Heidegger's theory of enframing, according to which social and "non-technical forces ... would be merely ontic and subordinated to the ontological fundamentals revealed in the technical functionalization of the world."[3] According to Feenberg's interpretation, technical culture can be seen to work as a substitute for being and seems to have become the transcendental source of all meaning. His question is: how is this meaning constituted and how can it be democratized?[4]

By focusing on some of Feenberg's most influential works, such as *Questioning Technology* (1999), *Heidegger and Marcuse: The Catastrophe and Redemption of History* (2005) and *Between Reason and Experience: Essays in Technology and Modernity* (2010), I attempt to show how Feenberg's productivist interpretation of Heidegger's early phenomenology connects with his positive critique of the later Heidegger's concept of the technological enframing. While Heidegger advocated an abstractly conceived, intermittent bracketing of technical use that moved away from technical devices, Feenberg takes an opposite tack. He develops a two-tiered theory regarding the function and meaning of technical devices that pays attention to their potentialities. Combining this theory with Marcuse's call for an "aesthetic *Lebenswelt*" helps Feenberg to articulate a phenomenological rationality based on a technologically reformed social order that strives for the materialization of life-affirming aesthetic and political values.

In his writings from the late 1990s to the present, Feenberg's body of work underscores the presence of two Heideggers, i.e., the early Heidegger whose ontological phenomenology of everyday life can be seen to affirm the actualization of technical potential and the later Heidegger, whose phenomenology of technology points to a world that is reduced to fungible, raw material. Feenberg arrives at his interpretation of Heidegger's early phenomenology, as a phenomenology of production modeled on craft work, by working through what he interprets as Heidegger's dystopian ontological account of the technical world as Heidegger describes it in his now famous essay "The Question Concerning Technology." There is also a third Heideggerian influence on Feenberg's phenomenology of technical action. This influence is Heidegger's critique of substance metaphysics as described in his collection of writings from the late 1950s and 1960s in essays such as "The Thing," "Building Dwelling Thinking," and "What are Poets For?" These essays highlight the world disclosing power of local things and practices while also lamenting the world-destroying effect of technical devices. With the help of Heidegger's critique of substance metaphysics, Feenberg transposes Heidegger's analysis of the thing, as a situated gathering that discloses local worlds, onto technical action and technical devices.

In the first part, I will turn to the influence that Heidegger's concept of the technological enframing exerts on Feenberg's philosophy of technology. In the second part, I will discuss the influence of Heidegger's early phenomenology on Feenberg's phenomenology of technical action. These two influences, I argue, converge in Feenberg's instrumentalization theory, later rearticulated as the functionalization theory of technology. Partially inspired by social constructivism, an approach that "emphasizes the social contingency of technical development,"[5] the instrumentalization theory can be seen to provide much needed content to Heidegger's theory of technology. Furthermore, the theory sets itself apart from the dystopian critical theories of Marcuse and Adorno that arrive at a "totally administered" society; from theories that identify a separation between society and technology, such as Habermas'; and finally, from radical constructivist theories that simply collapse the difference between society and technology.

The Enduring Influence of "the Question Concerning Technology"

Heidegger's ontologization of the technological world put modern technology on the philosophical map. His theory identified the world as technological by equating efficiency with technology, identifying technology with the modern world and then defining this world as one that is governed by the imperative of efficiency. In "The Question Concerning Technology" Heidegger uses the neologism enframing (*Ge-stell*) to refer to a Western attitude of calculation and imposition, or "challenging-forth" (*Herausforderung*), that treats nature and people as fungible raw material, disclosing them as available resources or "standing-reserve."[6] According to Heidegger, the essence of technology is "nothing technological," nothing on the order of the machine, although its explanatory reach includes machines and non-machines alike. Enframing is not a dystopia or a widespread problem we could solve with appropriate remedies, but the underlying structure of being in our time, a historical "mode of revealing" that foregrounds dominant norms such as efficiency, calculability, control, and ordering, while concealing others. It is ontological rather than ontic (an object of choice), to use the terminology Heidegger applied in his earlier work, *Being and Time*. Heidegger claims that to show up as a resource of any kind is always already to be thrown (*geworfen*) into an orderable world, to be made available, to control or to be controlled.

Heidegger's concept of the enframing innovate a critique of the mainstream, instrumental understanding of technology that sees technology as a neutral tool. This instrumental attitude is "correct but not true" as it is "limited to a subjective set of conditions"[7] that it presupposes but cannot explain. One such (significant) subjective condition is the still dominant, seventeenth century Cartesian and mechanistic assumption that individual rational subjects control objects, i.e., those entities perceived to lack reason and agency. This mechanistic assumption continues to promote a widespread instrumental attitude still popular today.

However, since instrumentalism is stuck in modern subjectivism, it is unable to recognize its own historical bias. For example, it cannot tell us *why* rationality, objectivity, and maximizing output matter but simply tells us *that* they matter. It cannot explain why today, for instance, norms such as efficiency, order (ability), and control are dominant and not others, such as tranquility, piety or spontaneity. Moreover, the rational and instrumentally minded subject fails to notice that what she does not master is the "will to mastery" and that in multiple contexts s/he now treats herself as an object, that is, she has become a resource.

In an earlier version of "The Question Concerning Technology" entitled "Positionality," Heidegger describes the resource as being under "attack" and "forced into conscription."[8] Whom the attack benefits or harms, pleases or pains, could be relegated to variable, subjective factors, but the fact remains that being a resource at all involves sustaining one kind of "attack" or another. Implicit in this description of violence is a condemnation of violence, a value judgment coiled at the heart of an ontological theory. Heidegger calls *Gestell* a "plundering" (*Geraff*)[9] that shows up as a kind of human "exploitation" or drive for "spoils and profit"[10]; yet this does not address its ontological dimension.

In "The Question Concerning Technology" Heidegger names "that challenging claim which gathers man thither to order the self-revealing as standing-reserve: "*Ge-stell.*" Its "chief characteristics" are to expedite, regulate, and secure *for its own sake.*[11] In his later phenomenology (broadly construed), he explains "resource" briefly and in generalities, for example, as "the way everything presences," "whatever stands by in the sense of standing-reserve no longer stands over against us as objects,"[12] and through terse phenomenological examples.

In the earlier "*Das Ge-Stell,*" however, his account of the resource is more elaborate, pointing to the (re)ordering of the world into an inventory of stuff that is uniform, equivalent, interchangeable, uprooted and made available upon request. He says: "The standing reserve is much more that which has been shattered [*Zerstückte*] into the orderable. This shattering does not break apart, but instead precisely creates the standing reserve of the pieces of inventory."[13] In "Science and Reflection" he claims that the "subject-object relation reaches, for the first time, its pure 'relational,' i.e., ordering, character in which both the subject and the object are sucked up as standing-reserve." Paradoxically, this "does not mean that the subject-object relation vanishes but rather the opposite: it now attains to its most extreme dominance… a standing reserve to be commanded and set in order."[14]

Terse phenomenological notes about the transformation of the Rhine River into a mere resource for the electrical power plant that suppresses the folkloric dimension of the landscape point to the resource character of the river. According to Heidegger's famous phenomenological analysis of the power plant on the Rhine, the river no longer gathers local traditions and lore, but is primarily seen as a water resource *for* the power plant. It is built into the dam, rather than the dam being built into the river. In addition, his phenomenological interpretations of the tourist industry and the patient supply for a clinic,[15] of foresters and radio listeners,[16] describe a seamless subordination of subjects to objects where, for example, tourists are ordered to special vacation sites and radio listeners become optimizable fodder for the "the public sphere." There is a deeply normative aspect to this ordering "as attack" that paradoxically manifests as lacking coercion and not producing distress. For instance, tourists are not angry at being told what to do. Instead they follow directions well and appreciate their vacations being optimized. Radio listeners enjoy music and news. Most are not concerned about becoming cogs in the public opinion machine, or that "a requisitioning and positioning … has intervened in the[ir] human essence."[17] Yet Heidegger insists that everywhere a lack of distress is *the real distress*, rendering "the human condition, man's being, tolerable for everybody and happy in all respects."[18]

In "the world of technology" constant ordering as "continuous attack and forcing into conscription" discloses a world of uniform, equivalent and replaceable entities[19] whose potential is "stored up… distributed… and switched about ever anew," without serving a fixed and final goal, without limit (*peras*). Almost ten years later, in "Traditional and Technological Language," he underscores this idea as follows: "What is peculiar to technology resides in … the demand to challenge nature forth into placing it at our disposal as securing it as natural energy. This demand is more powerful than any human positing of ends."[20] Human agency alone cannot overpower this demand, nor can individuals choose their way out of the enframing. Heidegger alludes to an overcoming of the technological age that may come as a result of deep acts of reflection and, as he says in his final interview with "Der Spiegel" (1966), through the proper attunement to the arrival of a "new god." These elusive proposals appeal to non-technological interventions and fail to persuade socially minded philosophers, such as Feenberg.

Overall, Feenberg rejects the Heideggerian enframing and does not consider it to be an adequate explanation of technical production. He groups this theory with other early twentieth century, abstract, deterministic and dystopian, theories of technology that frame technology as an autonomous force, ready to sweep us up into the future. However, with a view to particular technologies, Feenberg agrees with Heidegger's assessment that enframing is *sometimes* a relevant feature. He writes

> Although Heidegger means his critique to cut deeper than any social or historical fact about our times, it is by no means irrelevant to a modern world armed with nuclear weapons and controlled by vast technically based organizations. These latter in particular illustrate the concept of the enframing with striking clarity. Alain Gras explores the inexorable growth of such macro-systems as the electric power and airline industries (Gras 1993). As they apply ever more powerful technologies, absorb more and more of their environment, and plan ever further into the future, they effectively escape human control and indeed human purpose. Macro-systems take on what Thomas Hughes calls "momentum," a quasi-deterministic power to perpetuate themselves and to force other institutions to conform to their requirements (Hughes 1987). Here we can give a clear empirical content to the concept of enframing.[21]

Yet the explanatory reach of the Heideggerian enframing is ultimately limited. Its imperative of efficiency does not succeed in summing up the form of modern life. Where Heidegger fails, Feenberg can be seen to argue, Heidegger's student and Feenberg's teacher, Herbert Marcuse succeeds much better. Marcuse's concept of *technological rationality*, as developed in his magnum opus *One-Dimensional Man*, provides a social and political ontology for rethinking the technological subject in Western, capitalist societies. Drawing on Heidegger and Marcuse,[22] Feenberg argues for a social ontology of technology that combines historical ontology with social concerns. He develops a two-tiered theory of technology that he calls the instrumentalization theory of technology. As we shall see in the third section, this theory does not entirely reject the Heideggerian criterion of efficiency but recasts it as merely a partial explanation of functionality, one criterion among many others that distinguish technological devices from other things.[23]

A Phenomenology of Production: Rethinking Potentiality in the Technical World

Feenberg's retrieval of the early Heidegger rests on Heidegger's reading of Aristotle. This reading claims, according to Feenberg's interpretation of Heidegger's main works on this topic, that is, *Aristotle's Metaphysics Theta 1–3: On the Essence and Actuality of Force* and "On the Essence and Concept of *Physis* in Aristotle's Physics B, 1," that technology serves as the never quite acknowledged (perhaps actively suppressed) model for being.[24] In Feenberg's influential book, *Heidegger and Marcuse: The Catastrophe and Redemption of History*, he uses Heideggger against Heidegger by wresting a phenomenology of production from a Heidegger that attempts to distance himself from a philosophy of making. Moreover, by reading Heidegger through the philosophy of Marcuse's social dialectic, Feenberg draws out the ethically normative implications in Heidegger's critique of instrumentalism. Feenberg's goal is to show that technology is always already ethically and politically structured; its neutrality is merely ideological. He argues that a phenomenological interpretation of technical devices reveals that their meaning is socially transformable insofar as both their production and implementation are always already conditioned by an inherited historical horizon. The task is to recognize "their inherent possibilities" based on their practical integration into the lifeworld. Like all things, technical devices,

> ... do not have fixed essences separate from their manifestations because they are not themselves stable and fixed. Rather, they belong to a field of interactions which establishes their inner coherence and their boundaries... Potentialities are inscribed in things but do not constitute them as independent Aristotelian substances. Instead, something in the constellation of their present connections gives a direction to their development.[25]

Although the later Heidegger tries to distance himself from understanding nature (*physis*) on the model of *techne* as making, Feenberg brings out the significance of a suppressed theory of *techne* in Heidegger's works. He insists that Heidegger's phenomenology remains indebted to the Greek understanding of being (*physis*) on the model of *techne* where the presence of both natural and artificial beings is understood on the model of production. But in *Being and Time* Heidegger gives this approach an original formulation. Here meaning emerges as the relationship of *Dasein's* circumspective concern rather than somehow preexisting it, or the converse, originating in subjective cognition. According to Feenberg,

> The production model continues to operate but in a phenomenological context. Heidegger presents everyday *Dasein* as primarily handling objects, i.e., involved in *techné*. But the equipmental realm, now defined as the "world," is no longer approached through the structure of the product as it is by Aristotle. Instead, Heidegger develops a phenomenology of

> *Dasein's* use of objects [as *readiness to hand* and/or *presence at hand*]. This approach grounds the *eidos* in *Dasein's* temporality.[26]

This phenomenological conception of essence reconciles Greek objectivism with modern subjectivism and overcomes the contradiction between objectively given essences and subjectively chosen plans or goals. For existential ontology, "The essences of things are no longer reified as things themselves but understood from out of their place in everyday practical activity," phenomenologically interpreted. Paradoxically, this new understanding of essence is made possible by "modern technology."[27] The rise of mechanical production is connected with the breakdown in the concept of essence.

According to Feenberg's Heidegger, the Greeks believed that essences, that what endures, the *eide*, are not arbitrary products of the human will "but arise from being itself." The limitation of the Greeks was to view the world in terms of discovery rather than revealing, and this discovery itself reduced the self-showing (*phainesthai*) of being to immutable *eide*, rather than grasping it historically. Heidegger historicizes[28] the self-showing of being and so provides the corrective to the ahistorical, Greek understanding of being. Feenberg is both attracted and repelled by Heidegger's ontological sense of history. He is concerned that an ontological sense of history relegates human agency to the sidelines of social, historical, and more importantly, technological participation. Nonetheless, inspired by Heidegger he attempts a (neo-Hegelian) dialectical reconciliation of modern and ancient understandings of *techne* that respects the phenomenological given while at the same time expanding the role of subjective creativity.

Thus, Feenberg argues that while the Greeks failed to acknowledge the historical contingency in their understanding of essence and "confounded being with the essences of particular things," the moderns "confused the essential role of humanity in the process of revealing with technical command of nature"[29] through the arbitrary imposition of the will. He strives to articulate a synthesis between antiquity and modernity, in "a relation of human being to being that can replace both Greek bringing-forth and the modern *Gestell*."[30] That is, he applies a Hegelian remedy to a Heideggerian malady. This application relies heavily on Feenberg's acceptance of Marcuse's critical recasting of Heidegger's *Gestell* as *technological rationality*.

Following Marcuse's writings, in *One-Dimensional Man* and elsewhere, Feenberg believes that creative subjectivity is ruined by the systematic rationalization of irrational, because life-denying, desires and choices that are dominant in Western, capitalist democracies. Yet, since this domination is socially determined it is also socially transformable through an aesthetic revolution, for Marcuse, and through the active incorporation of "participant interests" in the historical development and implementation of technology, for Feenberg. However, these possible transformations are foreclosed in the totally administered system of the Heideggerian *Gestell.* It is precisely because Feenberg collates the postmodern resource with the modern subject, in Heidegger's account of the enframing, that he is able to, with the help of Marcuse's political ontology, resuscitate a subjective creativity from within the de-worlded technological world.[31]

In Marcuse's work, Feenberg sees an alternative that combines Heidegger's analysis of *Dasein's* factical life with Marxist politics to build concrete forms of resistance, first and foremost, to the ontological positivity of technological rationality.

He writes:

> Following Lukacs, Marcuse interprets … politics in terms of the struggle against reification, reification cast in the role of inauthentic objectivism in Heidegger. The equivalent of authenticity now appears a solidarity in that struggle. All Marcuse's later attempts to reach the concrete, through such concepts as the "new sensibility," sensuousness, the aesthetic, the instincts, resonate with this original existential Marxist approach.[32]

However, in a one-dimensional society that forecloses the possibility of real opposition, the recognition of new (and non-exploitative) possibilities as they apply to both humans and nonhumans, demands the prior liberation of the individual at the most fundamental and non-reducible level, that of experience itself, a level where reason and imagination are linked, where reason incorporates the imagination rather than defines itself against it, as in Cartesian inspired metaphysics and politics.

For Feenberg the goal is to make this link in the technical base explicit in order to further a democratically constructed technical world that will promote what Marcuse called "the pacification of existence" rather than a continued capitalist "struggle for existence." From Heidegger, Feenberg adapts a phenomenology of technical action that combats the ideology of the neutrality of technology. From Marcuse, he learns about the real dangers of the ideology of the neutrality of technology and the importance of engaging the imagination to combat this seemingly rational ideology.

The Lifeworld of Technical Things

The previous sections have engaged the influence of the early and late Heidegger on Andrew Feenberg's philosophy of technology along with his response to these influences. In this section, I will attempt to bring them together, as Feenberg does, in his rational critique of (social) rationality summed up in "Instrumentalization Theory of Technology." This critique is discussed at length in *Questioning Technology* (1999) and recast as "The Functionalization Theory of Technology" in his "The Concept of Function in The Critical Theory of Technology" (2016).[33] I see this theory as showcasing Feenberg's contribution to a threefold goal, that is, the undermining of technological determinism, the overcoming of a reified division between society and technology, and the historicizing of the life of technical things.

As I have already indicated, Feenberg consistently argues for a theory of technology that approaches technology from *within*, that is, that follows its historical development along socio-political and economic lines. Like Heidegger, he rejects the instrumentalist view that casts technology as essentially neutral, a mere device to which a social function and/or value is later added. Yet unlike his reading of Heidegger, he is not a technological determinist. His instrumentalization theory can be seen as a disaggregation of the concept of *Gestell*, bracketing out the unique content[34] of Heidegger's theory that, *pace* Heidegger, reveals the world disclosing, social power of technological devices.[35]

At the heart of Feenberg's phenomenology is the attempt to show how technical *design* is always already socially encoded. He strives to revive the role of human agency in the manufacture and adaptation of technologies, through democratic means. To this end he develops an anti-essentialist theory of technology that shows the inextricable connection between the functionalist character of technologies and their implementation or realization. His instrumentalization theory critiques the traditional privilege accorded to function, especially its reduction to efficiency, of technical devices. Heidegger is among the targets of this critique as his theory is guilty of reducing the device to mere efficiency.[36]

Feenberg argues that devices cannot be reduced to an abstract and non-relational function that is independent of social concerns and participant input. The design of particular technologies always already incorporates social concerns and these are not merely added afterwards. Uncovering the social dimensions of technologies helps to see them as multi-stable entities and open systems that interface dynamically with their users in various contexts. Feenberg shows that we can solicit the potential and power of devices once we acknowledge and actively commit to active participation in their development. That is, once we recognize their belonging to the lifeworld. He writes:

> To reduce technology to a device and the device to the laws of its operation is somehow obvious, but it is a typical fallacy of *misplaced concreteness*. Just as the parts of the clockwork mechanism lack true independence as such, even though they can be disassembled and identified as separate things, so technologies are not truly independent of the social world. That world is not merely an external environment; it traverses them with meaning.[37]

Feenberg wonders why Heidegger refuses to extend his practical analysis of the thing to the device. He is inspired by this *lacunae* in Heidegger's meditation on the gathering power of the thing as explained in his essays "The Thing" and "Building Dwelling Thinking." Heidegger's insights undermine the understanding of persons and things along the lines of substantive metaphysics, as anonymous variables, abstracted from their immediate environment or lifeworld. According to Heidegger's "Building Dwelling Thinking"

> Our thinking has of course long been accustomed to understating the nature of the thing. The consequence, in the course of Western thought, has been that the thing is represented as an unknown X to which perceptible properties are attached. From this point of view, everything that already belongs to the gathering nature of this thing… appear[s] as something that is afterward read into it.[38]

In *Questioning Technology*, Feenberg claims that "Heidegger grasps the thing not just as a focus of practical rituals, but as *essentially* that, as constituted qua thing by these involvements rather than as preexisting them somehow and acquiring them later."[39] Feenberg enlists Heidegger's relational and localized ontology of things to make sense of the life of technical devices and systems, to historicize them and at the same time to empower their users, the multitude of participants continuously propelling technical networks forward.

According to Feenberg, an engaged reflection with the development of devices shows them to be world disclosing sites of local gathering based on shared interests and practices. Heidegger assigns world disclosing power to works of art such as the Greek temples and later to local artifacts and architecture, such as jugs and bridges. But Feenberg protests. He asks, "If a Greek temple can open a space for the city, why not a modern structure? At what point in its development does architecture cease to be 'art' and become technology?"[40] Heidegger is unable to answer this question because he does not think concretely about technology and so fails to consider the developmental dimension of devices. Moreover, because Heidegger insists on evaluating technology from *without* by adopting a substantivist stand (with respect to technology but not with respect to things and persons!), he cannot extend what could be considered as his reflective and participatory analysis of the thing to the device. Thus, he reproduces the position of the technocrat or the "system manager" who reifies the device by thinking that he can predetermine its function without considering the socio-political demands of the lifeworld. But time and again the device has morphed beyond the initial scope assigned to it and this change in design has happened in direct response to socio-ethical concern, what Feenberg deems as the secondary level in his instrumentalization theory.

By using Heidegger's meditation on the world-shaping role of the thing against Heidegger's critique of the deworlding effect of technical devices, Feenberg's work highlights a significant inconsistency in Heidegger's critique of substance metaphysics. This insight allows Feenberg to extend Heidegger's analysis of the relational essence of the thing, as field or network, to the device. Thus, the analysis of the thing *as a network* can be transposed onto the technical device so that the latter is now grasped from within the practical standpoint. The world only reveals itself as such

> to a reflection that knows how to get behind cognition to a more primordial encounter with being. Such phenomenological reflection places us inside the flux of significance in which the world as network consists. This is not a collection of objective things, substance, but a lifeworld in which we actively participate and which only comes to light insofar as we understand participation as the most fundamental relation to reality.[41]

According to Feenberg, the meaning and role of devices emerge in relationship to a lifeworld, a background of social practices and meanings from which they arise and which they help to invisibly shape. Devices are not fixed, neutral things or tools that accrue meaning after they are made. Rather, they are always already carriers of social values whose meanings change in relationship to different users. For example, Feenberg claims that the size of a common household appliance such as the refrigerator depends on concrete social relations between men and women, so that

> where shopping is done daily on foot, refrigerators tend to be smaller than where shopping is done weekly by automobile. The technical design of this artefact depends on the social design of society…The refrigerator reduces the time spent shopping in a context in which household labor is still signified as largely feminine… The refrigerator is thus not just a labor saving device but, more concretely, it is a substitute for female labor in the nuclear family.[42]

This example, and others that I will use below, illustrates that analyses of technical devices reveal whose interests the devices always already (gather and) serve, and whose interests they frustrate. In this sense, devices can be said to point to a society's "technological unconscious." In sum, a close phenomenological reflection on particular technologies brings out their gathering and disclosive potential once we are able to see the two-fold depth of technical production. In his instrumentalization theory, this depth includes a primary and secondary level that correspond to "objectifying" and "subjectifying" moments of production.

In *Questioning Technology* Feenberg claims that his two-level theory includes not just the "functional constitution of technical objects and subjects" but also their actual place in the lifeworld as the "realization of the constituted objects and subjects in actual networks and devices"[43] that is, "primary instrumentalization" and "secondary instrumentalization" respectively. "Primary instrumentalization" is the only level considered by determinists, and it consists of "decontextualization, reductionism, autonomization and positioning," while the secondary level, as the social level of technical realization, consists of "systematization, mediation, vocation and initiative."[44] Secondary instrumentalization allows for the vital integration and re-integration "of technologies to larger technical systems and nature, and to the symbolic orders of ethics and aesthetics, as well as their relation to the life and learning processes of workers and users, and the social organization of work and use."[45] This twofold ontological structure effects and is in turn effected by the ontic changes in the design of particular technologies.

Feenberg underscores his theoretical insights with concrete examples. These examples expand the constructivist understanding of technology that interprets the development of technical devices as a "fit between devices and the interests and beliefs of the various social groups that influence the design process. What singles out an artifact is its relationship to the social environment, not some intrinsic property."[46] For instance, the bicycle, as we know it today, was once an ambivalent structure. Feenberg claims, "The object that we take to be a self-evident 'black box' actually started out as two very different devices, a sportsman's racer and a means of transportation."[47] The sportsman bicycle had a larger front wheel that favored speed at the expense of safety. The layman's bicycle had equal-sized wheels that compromised speed but enhanced safety. Eventually the social concern with safety mainstreamed the equal-sized wheel bicycle and marginalized the other. Here we see how the design (the ontic dimension) is informed by primary instrumentalization, the function, and secondary instrumentalization, ethical concerns, at the same time. The two are inextricably connected in experience. Feenberg points out that the gathering and disclosive power of the device stands or falls with the acknowledgment of "secondary instrumentalization" as part of the fundamental structure of the device.

In an earlier work, *Alternative Modernity*, Feenberg already argues for the gathering power of the Internet in the form of online patient groups such as the "Prodigy Medical Support Bulletin Board devoted to ALS." This study reveals the way in which anonymity enabled people to come together and speak honestly about the debilitating effects of their disease. The function of these online groups hardly "challenged" the patients "forth," nor did it reduce them to mere resources for endless optimization, as Heidegger's analysis anticipated. Rather, "Prodigy discussion participants established a list of priorities they presented to the ALS Society of America" and showed that other "online patient meetings have the potential for changing the accessibility, the scale, and the speed of interaction of patient groups."[48]

We experience devices as alienating and harassing if we subscribe to their autonomous force and turn a blind eye to their social coding, contextualizations and possibilities for recontextualizations. This is precisely

Heidegger's position. While Heidegger's highway bridge counts as a "thing" that discloses a local world, Feenberg points out, in a similar example, that some bridges are rather exclusive. For instance, the expressway designed by Robert Moses to connect New York with the beaches on Long Island included overpasses that were too low for city buses. Several decades ago this design effectively excluded poor people, who did not own cars, from mingling with the affluent classes.[49] The example reveals the racist and classist norms inherent in the bridge's design and thus, in the culture at large. Feenberg's social phenomenology allows us to see the concreteness of the bridge's "gathering" as it is foregrounded by the cultural dimension of his instrumentalization theory. Thus, the bridge is not merely the sum total of its material production, its neutral *functionalization*, that is, transporting people to and from their beach destinations and its *positioning*, that is, the "control of the consumer through product design."[50] It also simultaneously embodies classist and racist social norms that unwittingly privilege architecture as a *vocation* in general and reinforce the artistic prestige of this particular architect.

Heidegger's abstract theory of enframing elides these intertwining dimensions of nature and culture, function and value, brought to light by Feenberg's social phenomenology of technology. Feenberg's work expresses a philosophical ambivalence toward the early and later works of Heidegger. He retrieves Heidegger's phenomenology from *Being and Time*, but insists it is modeled on a productivist understanding of Aristotelian *techne* as making, a reading of Aristotle and *Being and Time* at odds with Heidegger's. Furthermore, he turns the later Heidegger against himself when he invites us to consider the gathering and world disclosing power of technological devices, a power reserved by Heidegger for works of art and craft only. Finally, Feenberg's theory of instrumentalization can be seen, in part, as a unique convergence of Heideggerian phenomenology and Marcusean critical theory, that can help us to better understand the scope and politics of the technological age.

Epilogue

Feenberg's theory of "primary and secondary instrumentalization" is helpful for understanding the uses of gendered technologies, such as various forms of advanced reproductive technologies or ARTs. For example, in my forthcoming book *Heidegger, Reproductive Technology and*

The Motherless Age (Palgrave Macmillan, 2017), I transpose phenomenological aspects of Feenberg's instrumentalization theory to make sense of in vitro *fertilization* (IVF) and to critique the misguided, because overly functionalized and reductive interpretation of this procedure. For instance, I show that:

> Stepping up to a high tech (IVF) intervention reconfigures how the woman's body is medically revealed so that now the woman as a whole person and a potential mother is concealed and she becomes revealed to the medical gaze as a collection of malfunctioning reproductive parts in need of technical assessment and optimization.[51]

I argue that this technological intervention enframes the woman's reproductive body and I show how Feenberg's analysis of the decontextualization of the "object" helps to concretize the enframing. Furthermore, I argue that the phenomenological application of Feenberg's theory helps to reveal the sexist dimension of this medical intervention. In "Impure Reason" Feenberg writes, "To reconstitute natural objects as technical objects, they must be 'de-worlded,' artificially separated from the context in which they are originally found so as to be integrated to a technical system. The isolation of the object exposes it to a utilitarian evaluation."[52] The eggs are indeed isolated, or as Feenberg claims, they are technologically "reconstituted" in order to test their reproductive usefulness, the usefulness of the sperm or because testing is a desirable research goal in itself. Once separated from the womb, the eggs can be seen to "reveal themselves as containing technical schemas, potentials in human action systems which are made available by decontextualization."[53] They become frozen eggs or frozen embryos stored for future fertilization. The cryopreservation opens up possibilities for embryo research and experimentation often *unrelated* to reproduction.

By embedding Feenberg's instrumentalization theory in my reinterpretation of Heidegger's concept of the enframing, I argue that the removal and removability of the eggs from the womb reveals the woman as stock, an *interchangeable* collection of available reproductive parts and functions. This is a significant step in the control and orderability of human (re)production and an early hallmark of what I describe as "reproductive enframing."[54]

Feenberg's *decontextualization* is coupled with a second step, *reductionism,* where the natural object is reduced to its primary qualities, such as "size, weight, and shape" or anything else about the "object that offers an affordance."[55] In the case of the eggs, doctors seek high quality (functional) eggs that contain the proper chromosomes, are young enough and resilient enough to combine with sperm and energetic enough to split (or grow) after fertilization. The eggs are reduced to these primary qualities because those seem most conducive to technical production, that is, embryo fertilization, growth, and implantation. Whatever the secondary characteristics of the eggs may be, they remain undiscovered. Feenberg writes, "Secondary qualities are what remains, including those dimensions of the object, that may have been most significant in the course of its pretechnical history. The secondary qualities of the object contain its potential for self-development."[56] This includes a thing's (or organism's) habitat.[57] It is unclear what precisely the secondary qualities of extracted oocytes may be but it is well known that ovulation induction disturbs their "habitat," the conditions necessary for the timely release of the eggs as well as the successful (re)implantation of the fertilized egg.

Finally, the "reproductive enframing" in IVF is underscored by what Feenberg calls the process of *autonomization*. Autonomization refers to the interruption of the *reflexivity* of technical action so that the subject can affect the object of technical production without seemingly being conversely affected herself or being only "slightly and disproportionately" affected. For example, the autonomization of IVF is visible when the medical staff fails to care for the woman's hurt feelings and mental distress, that is, collapsed life project, after an IVF cycle fails. Applying Feenberg's concept of autonomization brings out the fungible status of the woman. By choosing to dismiss the patient's distress the medical industry promotes an administrative and "purely functional" relationship with the patients. This affords them a kind of immunity from the consequences of their actions and reveals the disposable status of the individual patient.

Since there is no such thing as IVF in-itself, each of the technical abstractions involved in Feenberg's theory of primary instrumentalization relates to the lifeworld described in his account of secondary instrumentalization. With the help of Feenberg's theory, I show how decontextualization, reduction, and autonomization in IVF loosely correspond to moments in the "secondary instrumentalization" process, such as commercial *recontextualization, mediation*, and *identity* formation. Because IVF is always already a social phenomenon secondary instrumentalization can be separated from its primary counterpart only analytically. Within the current medical practices of artificial reproduction, the reduced (fertilized) eggs must be reintroduced, i.e., recontextualized, in the living womb of a woman, who must then be successfully integrated into a network of medical and administrative protocols. Finally, as I mentioned earlier, the autonomization process refers to a lack of *reflexivity* on the part of the doctors, the medical staff, and the woman. I argue that Feenberg's theory applies to the patient herself, who, alongside the medical establishment, often takes a merely functional attitude toward her use of IVF. Her corresponding identity is then supplied by contemporary, advanced industrial societies as a medical profession and the consumer of a service, respectively.

Notes

1. Andrew Feenberg, *Between Reason and Experience* (New York: Routledge Press, 2010), 181.
2. Andrew Feenberg, "From the Critical Theory of Technology to the Rational Critique of Rationality" in *Social Epistemology: A Journal of Knowledge, Culture and Policy*, vol. 22, nr. 1, March 2008, 7–8, 20. (2008)
3. Ibid., 13.
4. Andrew Feenberg, *Heidegger and Marcuse: The Catastrophe and Redemption of Modernity* (New York: Routledge Press, 2005) 24–25.
5. Andrew Feenberg, "From the Critical Theory of Technology to the Rational Critique of Rationality" in *Social Epistemology: A Journal of Knowledge, Culture and Policy*, vol. 22, nr. 1 March 2008, 14, 17, 20–21, 26. This central theory is also taken up in "Impure Reason" in *Questioning Technology* (1999), "Aesthetic Redemption" in *Heidegger and Marcuse: The Catastrophe and Redemption of Modernity*, among others.
6. Martin Heidegger, "The Question Concerning Technology" in *The Question Concerning Technology and Other Essays*, trans. W. Lovitt, (Harper & Row Pubs., 1977), 16.
7. Heidegger, "The Question," 4–6.
8. Martin Heidegger, "Positionality" in *Bremen and Freiburg Lectures: Insight into That Which Is and Basic Principles of Thinking*, trans. A. Mitchell, (Bloomington: Indiana University Press, 2012), 26.
9. Ibid., 31.
10. Ibid., 28.
11. Martin Heidegger, "The Question," 15–16.

12. Ibid., 17.
13. Martin Heidegger, "Positionality" in *Bremen and Freiburg Lectures: Insight into That Which Is and Basic Principles of Thinking*, trans. A. Mitchell, (Bloomington: Indiana University Press, 2012), 34.
14. Martin Heidegger, "Science and Reflection" in *The Question Concerning Technology and Other Essays*, trans. W. Lovitt (New York: Harper & Row Pubs., 1977), 173.
15. Heidegger, "The Question," 16–18.
16. Martin Heidegger, "Positionality," 35–36.
17. Ibid., 37.
18. Martin Heidegger, "What are Poets For?" in *Poetry, Language, Thought*, trans. A. Hofstadter, (New York: Harper & Row Pubs., 1971), 153.
19. Martin Heidegger, "Positionality," 42 & 62.
20. Martin Heidegger, "Traditional Language, Technological Language" in *Journal of Philosophical Research* Vol. XXIII, trans. W. Torres Gregory, 1998, 138.
21. Andrew Feenberg, "Technology and Meaning" in *Questioning Technology*, (New York: Routledge Press, 1999), 185–186.
22. Feenberg's theory of instrumentalization is also heavily indebted to Simondon's theory by the same name and to Lukacs's concept of reification. Discussion of these influences, however, lie beyond the scope of this chapter.
23. Andrew Feenberg, "Impure Reason" in *Questioning Technology*, (Routledge Press, 1999), 202–207.
24. Andrew Feenberg, *Heidegger and Marcuse: The Catastrophe and Redemption of Modernity* (Routledge Press, 2005), 84.
25. Ibid., 18.
26. Ibid., 41, my insert. The reference is to Heidegger's well known distinction between *Zuhandenheit* and *Vorhandenheit* in *Being and Time* (1927), and it is intended to underscore the privilege he accords to the mode of involved, practical engagement with the world over and against the detached approach of the rational observer. See Sects. 15 and 16 in Div. I.
27. Feenberg, *Heidegger and Marcuse,* 36.
28. Heidegger distinguishes between *Geschichte* and *Historie,* a historical or ontological account in terms of modes of revealing or *epochs* or bracketings that make up the history of being (*Seinsgeschichte*) and the ontic or empirical historiographical accounts of social events, as early as *Being and Time* (1927). The later Heidegger introduces sharp and relatively discontinuous historical breaks between the different "modes" of historical interpretation or "revealings" available in the West. See Heidegger, M., [1989] (1999) *Contributions to Philosophy* (*From Enowning*), trans. Emad, P. & Maly, K. (Bloomington: Indiana University Press, 1999).

29. Feenberg, *Heidegger and Marcuse*, 44.
30. Ibid.
31. As a critical theorist, it is not surprising that Feenberg remains unpersuaded by the totalizing, meta-historical pretensions and ontologically dystopian forecast of the Heideggerian *Gestell*. He insists that any totalizing narrative, including Heidegger's, requires an extopian perspective to explain the narrative. Yet this privileged point of view undermines the fantasy of the abstractly conceived totality, revealing its socially contingent constitution. For more on this, see Belu & Feenberg's "Heidegger's Aporetic Ontology of Technology" in *Inquiry* (2010). A fuller account of the ontologically dystopian character of the *Gestell* would have to engage its relationship to Heidegger's account of the history of being (*Seinsgeschichte*) as a forgetting of the clearing (*Lichtungsvergessenheit*). Whether or not the *Gestell* is coherent outside of this relationship remains a controversial issue in Heidegger scholarship.
32. Feenberg, *Heidegger and Marcuse*, 93.
33. Andrew Feenberg, "The Concept of Function in Critical Theory of Technology." In this very recent work he explains that "the new terminology is intended to block the temptation to order the instrumentalizations temporally. However, the new terminology too can be misinterpreted. The distinction of causality and culture is not intended to exclude causal modes of thought from the domain of culture in the broadest sense, but only to recognize the differentiation in modes of thought and institutions that actually exists in all modern societies. No fancy theoretical footwork can obliterate this differentiation in fact."
34. I refer to an internal feature of *Gestell*. Another unique feature, and one that I consider to be external, is discussed by Thomans Sheehan in his *Making Sense of Heidegger* (2015, 281) as the problematic relationship between *Gestell* and the *Seinsgeschichte*.
35. For a discussion of the central features of the functionalization and systematization stages in the theory of instrumentalization, see Andrew Feenberg, "Impure Reason" in *Questioning Technology*, (New York: Routledge Press, 1999), 205–207.
36. Andrew Feenberg, "Technology and Meaning" in *Questioning Technology*, (New York: Routledge Press, 1999), 185–186.
37. Andrew Feenberg, "Impure Reason" in *Questioning Technology*, (New York: Routledge Press, 1999), 213.
38. Martin Heidegger, "Building Dwelling Thinking" in *Poetry, Language and Thought*, trans. A. Hofstadter, (New York: Harper & Row Pubs.), 153.
39. Andrew Feenberg, "Technology and Meaning" in *Questioning Technology*, (New York: Routledge Press, 1999).
40. Ibid., 197.

41. Ibid., 196.
42. Andrew Feenberg, "From Critical Theory of Technology to The Rational Critique of Rationality" in *Social Epistemology: A Journal of Knowledge, Culture and Policy*, vol. 22, nr. 1 March 2008, 15–16.
43. Andrew Feenberg, "Impure Reason," 203.
44. Ibid., 208.
45. Feenberg, "Technology and Meaning," 193.
46. Andrew Feenberg, "The Limits of Technical Rationality," in *Questioning Technology*, (New York: Routledge Press, 1999), 79.
47. Ibid.
48. Questioning Technology, 192.
49. Ibid., 80.
50. Andrew Feenberg, "Impure Reason," 204–205.
51. Dana S. Belu, *Heidegger, Reproductive Technology & The Motherless Age*. (New York: Palgrave Macmillan, 2017), ch. 3, "Enframing the Womb: A Phenomenological Interpretation of Artificial Conception and Surrogacy in the Motherless Age." (Belu 2017)
52. Feenberg, *Questioning Technology*, 204–5.
53. Ibid.
54. Ibid. In connection to this point see *Heidegger, Reproductive Technology & The Motherless Age* where I claim that: "Reproductive enframing sums up the manipulation of the womb's potential by casting it as separable from the woman's body with which it was traditionally regarded as forming a whole. This manipulation introduces a fragmented approach to conception, one that frames the womb as a collection of discrete and movable reproductive parts: ovaries, follicles, eggs, fallopian tubes, hormones, and so on. These parts are managed as stock, potential reproductive energy that is challenged forth for further medical research and experimentation." (28)
55. Feenberg, *Questioning Technology*, 202.
56. Ibid., 204.
57. In "Impure Reason" Feenberg provides the example of a tree whose secondary quality as "habitat" no longer nourishes and shelters numerous species of flora and fauna once it is reduced to its primary quality, i.e., round wood.

Bibliography

Belu, Dana S. *Heidegger, Reproductive Technology & The Motherless Age*. New York: Palgrave Macmillan, 2017.

Feenberg, Andrew. *Between Reason and Experience: Essays in Technology and Modernity*. Cambridge, MA: The MIT Press, 2010.

———. "From the Critical Theory of Rationality to the Rational Critique of Rationality." In *Social Epistemology: A Journal of Knowledge, Culture and Policy*, vol. 22, nr. 11, 2008.

———. *Heidegger and Marcuse: The Catastrophe and Redemption of History*. New York: Routledge, 2005.

———. *The Philosophy of Praxis: Marx, Lukács and the Frankfurt School*. New York: Verso, 2014.

———. *Questioning Technology*. New York: Routledge, 1999.

———. *Transforming Technology: A Critical Theory Revisited*. Oxford: Oxford U P, 2002.

Heidegger, Martin. *Being and Time*. Translated by J. Stambaugh. State University of New York Press, 1996 (1927).

———. *Bremen and Freiburg Lectures: Insight into That Which Is and Basic Principles of Thinking*. Translated by Andrew Mitchell. Bloomington: Indiana University Press, 2012 (1949).

———. *Discourse on Thinking*. Translated by John M. Anderson. Harper & Row Publishers, 1966.

———. *Gesamtausgabe Band 79*. Frankfurt: Vittorio Klosterman, 1949.

———. "Only a God Can Save Us." In Der Spiegel Interview with Martin Heidegger. Translated by John. D. Caputo & Maria P. Alter. *Philosophy Today*, XX, 1976.

———. *Pathmarks*. Translated by William McNeill. Cambridge University Press, 1999.

———. *Poetry, Language, Thought*. Edited by J. Glenn Gray. Translated by A. Hofstadter. New York: Harper & Row Publishers, 1971.

———. "The Question Concerning Technology." In *The Question Concerning Technology and Other Essays*. Translated by William Lovitt. San Francisco: Harper San Francisco, 1977.

———. "Traditional Language and Technological Language." In *Journal of Philosophical Research, vol. 23, 1998.*

Marcuse, Herbert. *One-Dimensional Man: Studies in the Ideology of Advanced Industrial Societies*. Boston: Beacon Press, 1964.

———. *An Essay on Liberation*. Boston: Beacon Press, 1969.

CHAPTER 10

Future Questions: Democracy and the New Converging Technologies

Andreas Michel

But in reality we are not Gods.
(Feenberg 2010, xix)
The third great project of humankind will be [...] to upgrade Homo sapiens into Homo deus.
(Harari, 46)

Andrew Feenberg's work is astounding in its breadth. His ever more nuanced critical theory of technology is nothing less than a theory of modernity and a philosophy of political praxis. He takes his arguments from at least four distinct disciplines: philosophy, social theory, political theory, and science and technology studies. Yet the grand narrative he weaves in texts such as *Between Reason and Experience*,[1] the collection of essays that is the principal focus of my paper, is the narrative of Western Marxism (Lukács and the Frankfurt school, mainly Marcuse). In addition, his approach is suffused with the thought of Max Weber and Martin Heidegger, wherever their concerns intersect with those of

A. Michel (✉)
Rose-Hulman Institute of Technology in Terre Haute, Terre Haute, IN, USA

D.P. Arnold and A. Michel (eds.), *Critical Theory and the Thought of Andrew Feenberg*, DOI 10.1007/978-3-319-57897-2_10

Western Marxism. It is therefore no exaggeration to say that Feenberg continues the German twentieth century tradition of ideology critique.

The most prominent element that Western Marxism has in common with Weber and Heidegger is the description of modernity as entailing a *disenchantment of the world*,[2] the twentieth century interpretation of the nineteenth century reaction to the industrial revolution and its effects. In different ways, to be sure, yet with a common agenda, Marx, Nietzsche, Weber, Simmel, Lukács, the Frankfurt School, as well as Heidegger and Marcuse, address the disenchantment of the world. Originating with Max Weber, disenchantment connotes—albeit under different names (such as alienation or reification, for example)—the idea of privation: Modern Western culture is found lacking in normative foundation. The culprit, for most of these thinkers, is to be seen in the functionalization of modern life and its break-up into different independent social subsystems, each governed by instrumental reason. Martin Heidegger developed perhaps the ultimate short-hand term for this state of affairs with his notion of *Ge-stell* (*enframing*), a made-up term which is supposed to gather within it all the purely instrumental ways and interactions that govern modern society. This characteristic distinguishes modernity from previous societies that were based on a set of substantial values and were thought to have a natural teleology. What Heidegger captures here is the nature, pervasiveness, and exclusivity of the technological mode of revealing as the dystopian essence of modernity.

It is at this point where Feenberg both continues and subverts the tradition. His principal desire is to show that this all-encompassing view of technology—especially in Heidegger's notion of *Ge-stell*—is a misreading, a misunderstanding of what technology is and how it functions. He articulates his position on different terrains—for example, through the analysis of concrete technological artifacts or, in more programmatic fashion, through the elaboration of an alternative theory of technology based on the writings of Marcuse. The principal goal of his writings is to demonstrate that a view of technology as deterministic, autonomous, or dystopian is mistaken; rather, technology is shot-through with human design; and human agency in the face of technology is not only possible but also has occurred throughout history. His critical theory of technology is designed to further the democratic understanding of technology, which in turn will serve to counter the modern sense of disenchantment. Feenberg's main target in this endeavor is Heidegger's notion of *Ge-stell*.

In many ways, I agree with Feenberg's revisions of Heidegger's principal term. I believe that Feenberg's arguments are most convincing when he looks at particular technical objects or systems (such as the development of the bicycle or Minitel)[3] and demonstrates how human agency and human values are part and parcel of the design, development, and ultimate function of technical devices. In this fashion, he takes the wind out of an overarching theory of technology as disenchantment.

My concerns lie with the new "converging technologies for improving human performance"[4]—such is the title of the NSF-funded research into the convergence of technologies on the nanoscale—and how they might impact Feenberg's approach to technology, his views of a democratic politics of technology, as well as his critical theory of technology. For these technologies are beginning to insert themselves into the human decision-making process. But if this is the case, does this not suggest the triumph of the *Ge-stell* as the horizon of modernity—in the sense that future devices and systems based on the new converging technologies may infiltrate our lives to a degree that we freely surrender to them a level of authority and control in "our" decisions? Thus, the question of the interpretive force of the *Ge-stell* returns, one more time, to the center of debate.

In order to explore this issue I will, in a first section, outline Feenberg's nuanced acceptance of Heidegger's views on technology as well as his forceful rejection of the dystopian character of the *Ge-stell*. I will then move on to a discussion of his theory of instrumentalization which makes use of Heidegger's phenomenological approach. In the process, I hope to identify the principal elements that allow Feenberg to nevertheless reject the socio-political implications of the *Ge-stell*. In the second section, I will present the main ideas of Yuval Noah Harari's *Homo Deus*.[5] Harari's text will allow me to revisit the notion of human agency as theorized in the *Ge-stell* and relate it to Feenberg's account.

Throughout, I will follow Feenberg's advice: "If there is something of value in Heidegger, as I believe there is, it can be extracted only by sacrificing fidelity to his doctrine. The way to get at this worthwhile contribution is critically, not just exegetically."[6] Feenberg is right, I think, to say that we must go beyond Heidegger. What is at issue, though, is the direction of the step beyond. For Feenberg, Heidegger's limitations lie in the dystopian character of the *Ge-stell*, which Feenberg counters via a critical theory of technology that offers the possibility of positive change through human agency. Yet, the step beyond Heidegger could also go in the opposite direction, that is, that Heidegger's own response to technological revealing, either in the form of *Gelassenheit* or of *Verwindung*, is being "outperformed" by the reality of the *Ge-stell* as it is on the verge of being implemented in our daily lives through the algorithmic recreation of our lives as a second nature.[7] For the progressive technological politics that Feenberg envisions to be successful, they need to contend with the new, properly postmodern and post-humanist technologies that threaten to make such politics a thing of the past.

Put differently, the question that drives this essay is if Feenberg's approach, which works well for the sociological understanding of technical devices, holds up when it comes to the other dimension of the *Ge-stell*, which, as every reader of *The Question Concerning Technology* knows "is itself nothing technological."[8] Rather, the real danger is to be seen in the "spirit" of modern technology which, unceasingly, aims to put mankind in control over nature through ever more refined formalized structures of rationality.

Heidegger's Fruitfulness

The first thing that strikes one when reading Feenberg is how much he takes over from Heidegger. For someone who is thoroughly opposed to a dystopian view of technology and who makes it clear that he wants to "break with Heidegger,"[9] the following statement seems to come as a surprise: "[Philosophy of technology's] vantage point has been occupied fruitfully by Heidegger."[10] Of course, we have to see what exactly this fruitfulness consists in; but it is important to note from the outset that rather than rejecting Heidegger's descriptions of modernity outright, Feenberg finds some basic truths in them. In this first part of my paper, I want to show what these involve and why Feenberg goes along with Heidegger—up to a certain point.

To begin with, Feenberg himself belongs to the heterogeneous tradition around *disenchantment* I mentioned above. It flourished mostly among representatives of Marxist and neo-Marxist critics of modernity who, along with Max Weber, saw modern man as enclosed in an "iron cage" of bureaucracy, where efficiency and functionality are the only values. The Marxist tradition referred to this state of affairs as alienation or, in Lukács's formulation, reification. Feenberg summarizes this tradition in the following manner:

> Lukács provides the link between Marx and the Frankfurt School. Works such as *Dialectic of Enlightenment* (Adorno and Horkheimer 1972) and

> *One-Dimensional Man* (Marcuse 1964) are often dismissed as irrationalist and romantic when in fact they intend a rational critique of a new object. That object, omnipresent technology, is based on calculation and optimization and shapes not just technical devices and social systems but also individual consciousness. Organizations, technologies, and culture are inextricably intertwined, each depending on the other for its design and indeed for its very existence. According to the Frankfurt School, advanced industrial society is "totally administered" as a bureaucratic technical system.
>
> This extremely negative view of modernity results from a dystopian overemphasis on the limits of agency in socially rational systems. As a result the Frankfurt School often serves as a left-wing version of Heidegger.[11]

As Feenberg points out, this formulation of the neo-Marxist analysis of modern society has a lot in common with Heidegger's description of the *Ge-stell*. The main difference is that, for Heidegger, omnipresent technology constitutes the one collective horizon of modern life. Modernity therefore is essentially to be understood as *technological revealing*; that is, the *Ge-stell* represents the essence of modernity. Feenberg is intrigued by this similarity between Frankfurt and Freiburg when it comes to the significance of technology—especially so in his teacher Herbert Marcuse, who had been a student of Heidegger's and whose texts bear the traces of Heidegger's influence. Feenberg follows these traces when, in the last chapter of *Between Reason and Experience*, he lays out his own theory of technology, which he calls, in honor of Marcuse, aestheticized technology. Why does Feenberg turn to Heidegger? Wherein consists the fruitfulness of Heidegger for Feenberg's own critical philosophy of technology?

The answer is that both Marcuse and Feenberg are intrigued by Heidegger's phenomenological approach to knowledge, which he refers to as the revealing of being.[12] And within this account they are intrigued by Heidegger's analysis of the intimate relationship between art and technology (*techne*).

Heidegger's philosophy represents a radical critique of epistemology.[13] For Heidegger, people and objects come to be what they are in a pre-established horizon that he calls "world" and which provides the framework for their encounter with one another. Since this "world" frames the encounter, people and things come to appear as what they are in terms of this horizon. In Heidegger's language, people and things are *revealed* or *disclosed* in and through this "world," rather than as "horizonless" objects for subjects, as is the case in traditional epistemological accounts of knowledge. It is these concepts of *disclosure* and *world* that Feenberg finds useful.

The language of revealing comes to bear upon the topic that interests us here. One of the most influential of Heidegger's analyses is that of the meaning of the Greek word *techne*, which is at the heart of his understanding of both art and technology. According to Heidegger, in premodern times technical artifacts emerged in undifferentiated craft practice (*poiesis*) where the artifact represented a functional, aesthetic, and ethical whole. Differentiated modernity has made this emergence of art/technology a thing of the past. In the following passage, Feenberg summarizes the important distinction between premodern and modern technical/artistic practice:

> Greek techne is an undifferentiated practice. The meanings that underlie it are fixed by the culture so securely that they are not modified or questioned. These meanings are not strictly functional in our modern sense but include what we would call "aesthetic" and "ethical" values as well as technical considerations. The Greeks invented a philosophical terminology in which to refer to the complex meaning in which all these considerations are united, calling it the "essence" of the thing.[14]

For Feenberg's critical theory of technology Heidegger's interpretation of premodern craft proves to be decisive. In premodern practice, "craft, Heidegger argues, is a way in which things become what they truly are."[15] "In the technological revealing" on the other hand, "no essences are uncovered."[16] Feenberg is intrigued by the idea of an undifferentiated practice able to gather and express the essence of things in a functional, aesthetic, and ethical totality. "Without it there can be no critical reason."[17] Only as long as there is this connection is there a chance in modernity to overcome the dystopian aspects of the *Ge-stell*. Interestingly, it is Heidegger's theory of premodern craft, then, that provides Feenberg with a handle to counter Heidegger's own dystopian interpretation of the technological revealing as *Ge-stell*, which is characteristic of modernity. Thus, as much as he fights Heidegger's interpretation of *modern* technology, Feenberg adopts Heidegger's analysis of premodern *techne*, which is to guarantee the fullness of artistic artifact. In that sense, Heidegger can be seen to lay the groundwork for Feenberg's own more optimistic, alternative model of an aesthetic technology, as he develops it in the last pages of *Between Reason and Experience*, a model "that is neither premodern nor modern in the usual sense of the terms."[18]

To summarize my argument thus far: For Feenberg, the fruitfulness of Heidegger's thought on technology consists in Heidegger's phenomenological account of technology as the (potential) revealing of an undifferentiated essence, always understood not in terms of *essentia* but rather as bringing forth.[19] Feenberg subsequently uses this argument against Heidegger's dystopian view of *modern* technology where the access to such an essential revealing is in serious danger. In the remainder of this section I will point to what this approach enables Feenberg to do.

For this purpose, we need to look more closely at the principal notion in this context: that of essence. Like Heidegger, Feenberg does not conceive of essence in epistemological terms, that is, as representing the intrinsic nature of a thing. Rather, he follows Heidegger in conceiving of essence as the mode in which a total cultural artifact comes into being, combining functional, aesthetic, and ethical dimensions. In the essay entitled "Impure Reason" from his collection *Questioning Technology* Feenberg gives a sociological turn to these ideas by developing a "historical concept of essence." With this formulation he intends to capture the "sociocultural embeddedness" of technology: "The essence of technology [is] the systematic locus for the sociocultural variables that actually diversify its historical realizations." Put differently, Feenberg considers as the essence of technology all the "socially concrete stages" that go into the revealing of a technical object. "It is the logic of that process which will now play the role of the essence of technology."[20]

In order to capture the concreteness of the technical object as well as the technical process, Feenberg has developed a two-level instrumentalization theory. Primary instrumentalization refers to the "*functional constitution* of technical objects and subjects"—what Heidegger would call *Ge-stell*, which for him would be the exclusive mode of revealing in modernity. Feenberg however adds onto this a level of secondary instrumentalization, which is "focused on the *realization* of the constituted objects and subjects in actual networks and devices."[21] This level comprises the sociocultural context, as, for example, questions of aesthetics and ethics, besides others. Thus, the first level represents the general level of reification, while the second level of instrumentalization is designed to compensate for it.

Feenberg has put this reformulation of the essence of technology to work. In a number of case studies, for example, on the bike or on Minitel, he has successfully demonstrated that to look at technical objects from the point of view of their functional constitution alone cannot explain their design. It can be shown, on the contrary, that the level of secondary instrumentalization has a decisive influence on the design of technical objects. Furthermore, part and parcel of the level of secondary instrumentalization is the politics of technology:

> Diverse interests now contend for the influence over the design of technologies just as they have always fought for the influence over legislation. Each alternative design of medical technologies, transportation systems, the Internet, educational technology, and so on has its advocates whose ideology, way of life, or wealth depends on control of technical designs.[22]

With this statement we have arrived, I believe, at the heart of Feenberg's critical theory of technology: His desire is to describe technical objects and systems in such a way that they appear as the social and political reality they are—thus laying bare the potential of democratic participation in technology design.

From this vantage, then, it seems apposite to state that Feenberg extends the Enlightenment tradition to the field of modern technology. But unlike Enlightenment's disenchanters, Feenberg holds that (1) technology is not an autonomous field, but one of interests in contention that should be democratically adjudicated. Furthermore, (2) technology is not deterministic: Human beings can intercede in the production and design of technical objects. And therefore, (3) technology, including modern technology, is not dystopian, unlike what is implied by Heidegger's *Ge-stell*. Thus, based on close attention to the complex manner in which concrete technical objects gain purchase in society, Feenberg presents a positive, even optimistic view of technology without glossing over the primary, instrumental dimension.

> Real change will come not when we turn away from technology toward meaning but when we recognize the nature of our subordinate position in the technical systems that enroll us, and begin to intervene in the design process in the defense of the conditions of a meaningful life and a livable environment.[23]

This statement encapsulates the optimism of Feenberg's critical theory of technology. This optimism will be severely tested by a recent book on the future of our species, which, in its author's view, is potentially threatened by the latest developments in twenty-first century technology. This book sees the very humanism and liberalism that is the cornerstone of

Feenberg's critical theory of technology as imperiled. It therefore seems to resuscitate the ghost of *the Ge-stell*.

Upgrading Homo Sapiens

In Yuval Harari's recent book, *Homo Deus*, Heidegger's name does not come up; nor is there any reference to his theory of technology as a dystopian theory of modernity. Nevertheless, Harari's *Brief History of Tomorrow* (this is the book's subtitle) could not be more pertinent to the issues addressed in this article. Indeed, *Homo Deus* can be read, on one level, as a despairing update on Heidegger's description of the *Ge-stell*—in the sense that Heidegger's hope against hope for a free relationship to technology seems to have been all but dashed, and that modernity is in the process of succumbing to the dangers represented by the *Ge-stell*.

Harari's book is a three-part historical account of the Anthropocene, the "geological" age of mankind. He demonstrates how *homo sapiens* conquers the world (part 1), gives meaning to it (part 2), and finally loses control of it (part 3). As such, *Homo Deus* offers a grand sweep of human history in 400 pages, informed by in-depth knowledge of anthropology, world history, the history of science, as well as cultural, social, and political history. Harari has a gift for clearly and synthetically presenting his material and a sense of the big picture. In that respect, his work, like the work of Heidegger and Feenberg, offers a broad approach to a critical understanding of modernity. The book is full of fascinating and provocative detail surrounding the advances in deep machine learning and the creation of neural nets approximating the workings of the human brain—all of which would warrant extended discussion. However, with respect to the more narrow topic of my paper, I will focus on Harari's account of what he calls the "new human agenda." This challenges some of the results of Feenberg's critical theory of technology as discussed in the previous section. It challenges, in particular, the optimistic results of Feenberg's critique of Heidegger's dystopian vision as it manifests itself in the notion of *Ge-stell*. In challenging his optimism, Harari's text draws our attention again to Heidegger's views on the technological revealing in modernity.

While *Homo Deus* has highly speculative dimensions, it is not a futuristic—in the sense of science fiction—account of tomorrow since most of the technologies Harari investigates are already here, even if they have not all found widespread acceptance or application. Nor do the book's predictions necessarily paint a deterministic or dystopian picture of the future, even though they present a sobering one. In Harari's view, his predictions constitute "less of a prophecy and more a way of discussing our present choices. If the discussion makes us choose differently, so that the prediction is proven wrong, all the better."[24] What are his predictions? Harari maintains nothing less than that at present humanity is undergoing a radical shift from *homo sapiens* to *homo deus*—and, quite possibly, beyond. Humanity, in his view, is in the process of transcending itself, of turning an "information" version of itself first into a techno-human creature and then into an entirely new entity without biological characteristics.

> In the 21st century, humans will try to attain immortality, bliss, and divinity. This forecast isn't very original or far-sighted. It simply reflects the traditional ideals of liberal humanism. Since humanism has long sanctified the life, the emotions and the desires of human beings, it's hardly surprising that a humanist civilization will want to maximize human lifespan, human happiness and human power.[25]

The most recent technological advances make this new human agenda a real possibility. The convergence of technologies in the areas of artificial intelligence, bioengineering, cognitive science, and nanotechnology has led not only to unprecedented advances in these different fields; it has also given rise to a scientific, cultural, and political agenda that will leave *homo sapiens*, as we know him or her, behind. Today, the new technologies are hard at work at designing neural networks that aim to simulate and then upgrade the human brain. In the process, humanity might raise silicon-based intelligence to a position of power to which it might have to concede control in the long run.

Again, these reflections are not necessarily science fiction. They are based on the recent explosion of knowledge in artificial intelligence and its increasing importance in our lives. External algorithms have begun to tell us who we are and shape our daily interactions. They are increasingly relied upon in "our" decision-making processes. It is not hard to see these developments as manifestation of the essence of the *Ge-stell* as Heidegger conceived of it. As we will see below, external algorithms are used today to increase human efficiency, and thus to improve human beings and their surroundings. In the twenty-first century, the central technological challenge is quickly becoming the upgrade of mankind and our objects. Already over a decade ago, the above-mentioned NSF-sponsored report on *Converging Technologies for Improving Human Performance* (2003) included the following passage emphasizing the interdependence of the four research areas:

> The integration and synergy of the four technologies (nano-bio-info-cogno) originate from the nanoscale, where the building blocks of matter are established. This picture symbolizes the confluence of technologies that now offers the promise of improving human lives in many ways, and the realignment of traditional disciplinary boundaries that will be needed to realize this potential. New and more direct pathways towards human goals are envisioned in working habits, in economic activity, and in the humanities.[26]

The targets of the new technologies are, in no uncertain terms, the building blocks of matter, the very material of which humanity and the rest of the world is made. While "improving human lives" is a comparatively tame expression when compared to "improving human performance," both point to the new human agenda. Here, in the form of Harari's summary at the end of his book, is the heart of the matter:

1. Science is converging on an all-encompassing dogma, which says that organisms are algorithms, and life is data processing.
2. Intelligence is decoupling from consciousness.
3. Non-conscious but highly intelligent algorithms may soon know us better than we know ourselves.[27]

While Harari takes an entire book to develop these points, I will try here to summarize the effects and the potential impact these new technologies might have on modernity. The real crux of the convergence of technologies is the idea that *organisms are algorithms* and that, in essence, *life is data processing*. This at least is the most recent state of knowledge, based upon which researchers and engineers in the life sciences are designing new technological devices for us. They hypothesize—with much success—that (1) "humans are algorithms that produce copies of themselves"[28]; (2) human sensations, emotions, and desires are conceived as biological algorithms "honed by millions of years of evolution"[29]; and (3) "humans are an assemblage of many different algorithms lacking a single inner voice or a single self."[30] "The algorithms constituting a human are not free. They are shaped by genes and environmental pressures and take decisions either deterministically or randomly—but not freely."[31] On this view, what makes us human can therefore be converted into information, which can then be controlled through algorithms. Our organic nature can thus first be simulated, and then, if so desired, improved through technical devices based on external algorithms. And it is this merger of artificial intelligence and the life sciences that is at the heart of the changes that are poised to overtake Homo sapiens. As Harari says: "We are about to face a flood of extremely useful devices, tools, and structures that make no allowance for the free will of individual humans. Can democracy, the free market, and human rights survive this flood?"[32]

Harari provides many examples of technologies that we already trust and provide with authority, such as those relating to medical care and personal health. We trust these devices because we know that they already understand either the symptoms or our bodies better than we do—because they are able to process vast amounts of information that we don't have access to or cannot control. If this is the case, and if we continue, through Facebook and other electronic media and gathering devices, to freely provide the external algorithms with the necessary data they need for their calculations, then those computerized and linked devices will know us better than we do ourselves. We will end up entrusting them with our decisions. Yet another stage is reached when, in deep machine learning, algorithms turn into black boxes because their output can no longer be predicted, verified, or controlled based on the original input. "The seed algorithm may initially be developed by humans, but as it grows, it follows its own path, going where no human has gone before—and where no human can follow."[33]

In the last two chapters, Harari describes the two techno-religions that follow from these developments: techno-humanism and dataism.[34] Techno-humanism wants to "create *homo deus*—a much superior human model. [It will] enjoy upgraded physical and mental abilities that will enable it to hold its own against the most sophisticated non-conscious algorithms. Since intelligence is decoupling from consciousness, and since non-conscious intelligence is developing at breakneck speed, humans must actively upgrade their minds if they want to stay in the game."[35] "Data religion argues that humans have completed their cosmic task, and they should now pass the torch on to entirely new entities."[36] "Dataism now gives humanists a taste of their own medicine, and tells them: 'Yes, God is a product of the human imagination, but human imagination in turn is the product of biochemical algorithms.'"[37]

At that point the *Ge-stell* has taken over. Technological revealing will be the only game in town. And Feenberg's wish to "intervene in the design process in the defense of the conditions of a meaningful life and a livable environment"[38] will prove to be impossible. Modernity will have gone afoul of "catching sight of what comes to presence in technology, instead of merely staring at the technological. So long as we represent technology as an instrument, we remain held fast in the will to master it. We press on past the essence of technology."[39] Harari's book suggests that we are in the process of pressing on past the essence of technology.

Experience Endangered

The preceding sections present two different scenarios for human agency in modernity. The pivotal notion in both approaches is that of *experience*. Feenberg places in experience all hope of resistance to the *Ge-stell*, the technological revealing that enlists us in its service. For Harari, in a more all-encompassing sense, experience is the basis of all authority in modernity; it is what distinguishes it from medieval times. For both thinkers, and this is where we can find common ground, experience is intimately related to the fate of democracy, to individual freedom, to conscious interaction with nature. Yet, in Harari's look to the future, it is precisely this notion of experience that is fundamentally endangered. The last section of this article hones in on the different ways in which the authors assess the hope/possibility of resistance to the *Ge-stell* along the lines of experience.

In Feenberg's critical theory of technology, "experience" manifests itself in the concrete, lived sentiment that can be shown to be part of the design of technological devices. It occurs alongside the purely technical (rational) aspect of devices:

> Technical creation involves interactions between reason and experience. Knowledge of nature is required to make a working device. This is the element of technical activity we think of as rational. But the device must function in a social world, and the lessons of experience in that world influence the design.[40]

This introductory paragraph from the preface to *Between Reason and Experience* emphasizes the two dimensions of the technical artifact that

Feenberg's theory of primary and secondary instrumentalization is designed to capture. It represents his way of countering Heidegger's dystopian view of modern technology. In order to make his point, however, Feenberg has to identify a point of resistance to the *Ge-stell*, and he finds it in an emphatic notion of experience that remains free of the *Ge-stell.* How exactly does he arrive at this? Here we need to pick up the exposition from section one and see how, based on a critical reading of Marcuse's notion of aestheticized technology, Feenberg reintroduces the sphere of values through the lessons of experience into the *Ge-stell.* Our analysis of the first section becomes relevant here since Feenberg (via Marcuse) uses Heidegger's analysis of the Greek revealing of the essence of the thing as a foil to articulate his own critical theory of technology.

In order to counter the *Ge-stell* then, Feenberg performs a creative reformulation of Marcuse's idea according to which the detrimental aspects of modern technology can be revitalized through a return to the Greek version of revealing. Such a return, it is hoped, will show that accessing the sphere of values is still an option for a contemporary theory of technology: Values gained in/through experience would become part of the artifact that so comes into being. Put differently, lived experience engenders the values that the technical devices draw on alongside the purely functional dimensions of instrumental rationality. In his attempt to retrieve the sphere of values, Feenberg points to recent work in contemporary technology studies. Such work, in his view, has shown that concrete technical objects are never the mere result of means but come with values attached: "We no longer believe that technology is value neutral. Rather, contemporary technology studies argue that technological design always incorporates values through the choices made between the many possible alternatives confronting the designers. Technologies are not mere means but shape an environment in terms of an implicit conception of human life. They are inherently political."[41] Technological devices are not value-free but rather value-laden. As a result, just as in the Greek model, modern (aestheticized) technology can also be conceived as helping to give birth to the essence of the thing.

But where in modernity is such an "implicit conception of human life" revealed? What modern values can be shown to impose themselves in and through technical devices? Here, Feenberg offers a general example: In the present, the values attached to technical devices and systems (ought to) come into play whenever "the limits of the human body and nature" are at stake.[42] Lived experience, in other words, shows up these limits. In summary, one might say that Feenberg, by reformulating Marcuse and Heidegger, is convinced that—contra the *Ge-stell*—he has succeeded in articulating a critical theory of technology that is able to incorporate the values of the lifeworld and to thus provide a space for human agency and the opportunity for democratic participation in modernity.

In *Homo Deus*, Yuval Harari offers a different understanding of experience, a notion which—for him, too—is pivotal. As historian of world history, Harari sticks to the tripartite division of human history into Antiquity, the Medieval World, and Modernity; and it is in modernity that experience becomes the cornerstone of scientific and political authority. Whereas a large proportion of mankind had been beholden to the authority of myths and Gods in ancient times and to the one God in medieval times, mankind frees itself from such metaphysics in the modern age. But with this liberation from metaphysics comes the loss of ultimate meaning and a loss of trust in a cosmic order. To compensate for this loss modern man, according to Harari, devised the "covenant between science and humanism."[43] From this moment on knowledge is founded either in empirical science or else in human experience; and it is as part of this covenant that experience gains its pivotal status. At the heart of this covenant is modernity's greatest accomplishment: humanism. In the chapter entitled "The Humanist Revolution," Harari paints in broad strokes three types of humanism: liberal, social, and evolutionary humanism. In terms of political philosophy, we would talk of liberalism, socialism, and fascism. For Harari, these different humanisms are all versions of the modern turn to experience, which is therefore at the core of Humanism.

> What exactly are 'experiences'? They are not empirical data. An experience is not made of atoms, molecules, proteins, or numbers. Rather, an experience is a subjective phenomenon that includes three main ingredients: sensations, emotions, and thoughts. At any particular moment my experience comprises everything I sense (heat, pleasure, tension, etc.), every emotion I feel (love, fear, anger, etc.) and whatever thoughts arise in my mind.
>
> And what is 'sensitivity'? It means two things. Firstly paying attention to my sensations, emotions, and thoughts. Secondly, allowing these sensations, emotions, and thoughts to influence me. [...] Experiences and sensitivity build up one another in a never-ending cycle.[44]

Sensations, emotions, and thoughts—that is what experience means; it is what makes and guides the modern individual. The Humanist

Revolution therefore culminated in this: "the real source of authority is my own feelings."[45] While liberal humanism pursued the cult of the individual, social humanism that of the party, and evolutionary humanism that of Social Darwinism, their liberation from metaphysical constraints also meant that these different humanist sects (Harari) saw themselves as the bearers of Progress. No longer was history mapped out in a sacred text. Human beings were now free to create history. The pursuit of progress and how to best go about achieving it defines the nineteenth and twentieth centuries. Calling the different humanist sects religions, Harari describes the twentieth century as the Humanist Wars of Religion—which in the end were won by liberal humanism.

Modernity is thus the triumphant history of human experience, which raised the individual to the position of highest authority. But liberalism's victories are now endangered because, with the rise of the new converging technologies, the very status of experience is in question. We might be on the verge of entering into a postliberal age. "Liberalism," Hariri says,

> can(not) rest on its laurels. [...] The triumphant liberal ideals are now pushing humankind to reach for immortality, bliss and divinity [...] Yet what the scientists are discovering and what the engineers are developing may unwittingly expose both the inherent flaws in the liberal world view and the blindness of customers and voters. When genetic engineering and artificial intelligence reveal their full potential, liberalism, democracy, and free markets might become as obsolete as flint knives, tape cassettes, Islam and communism.[46]

The discussion below of what Harari means by the "inherent flaws in the liberal world view and the blindness of customers and voters" will challenge the pillars of humanist experience. What defined the strength of liberal culture—the trust in experience: in sensations, emotions, and thoughts—might be in the process of being outsourced to external algorithms.

During the last two decades, evolutionary biology and brain science have unleashed a radical attack on the most cherished aspects of liberalism: namely, on the notions of free will and of the authentic self. Recent laboratory experiments with robot mice and humans wearing transcranial helmets strongly suggest that what we have up to now considered as free will might in fact be the *consequence* of firing neurons rather than their origin. "If by 'free will' you mean the ability to act according to your desires—then yes, humans have free will, and so do chimpanzees, dogs and parrots [...]. But the million-dollar question is not whether humans and parrots can act out their inner desires—the question is whether they *can choose their desires in the first place.*"[47] Commenting on an experiment where the subject was asked to press the right or the left switch in response to a stimulus, Harari states:

> The decision to press either the right or the left switch certainly reflected the person's choice yet it wasn't a *free* choice. In fact, our belief in free will results from faulty logic. When a biochemical chain reaction makes me desire to press the right switch I feel that I really want to press the right switch. And this is true. I really want to press it. Yet people erroneously jump to the conclusion that if I want to press it I *choose* to want to. This is of course false. I don't *choose* my desires, I only *feel* them, and act accordingly."[48]

Feeling, Harari suggests, is not the originary act upon which liberalism is built, but derivative of a material event in our brains that precedes our conscious action. "As scientists opened up the Sapiens black box, they discovered there neither soul, nor free will, nor 'self'—but only genes, hormones, and neurons that obey the same physical and chemical laws governing the rest of reality."[49] But if I don't have a soul, free will, or a self, then the bedrock of liberal humanism—and especially its most cherished notion: the unique individual—has been eroded.

These scientific results obtained in the life sciences are now being "translated into everyday technology, routine activities and economic structures"[50] via artificial intelligence. Based on the not yet entirely proven assumption that "organisms are algorithms," many human tasks that require intelligence—not consciousness—will in the foreseeable future be executed by non-conscious algorithms.

The really important aspect of these changes as far as experience is concerned is that the system might still need humans *but not individuals.* Once we accept that individuality is a fiction and see the benefits of conceiving ourselves as biochemical algorithms, we might decide that we prefer to be constantly guided by a network of electronic algorithms, since electronic algorithms have enormous advantages when compared to biochemical algorithms. These external algorithms are fed with data from all over the world, they are constantly updated, and thus offer a much broader range of information based on which we can make decisions. But what this means is that *authority will shift from individual humans to networked algorithms.*[51] It remains true that we "dividuals"

(Harari) will make the ultimate decision as to how to act in any particular situation; but the way in which we arrive at these decisions has very little to do with our experience and feelings. We will put our trust in numbers rather than in feelings because this is where authority will be seen to rest. Human experience will become part of an integrated network.

What are the political ramifications once we can no longer in good faith think of ourselves as individuals; once we let our judgments be guided by external algorithms because we trust them more than our feelings? Over time, Harari suggests, the algorithm might turn from oracle to agent and then to sovereign. He provides the example of a GPS system that is constantly being fed in real time by millions of users. At first, as oracle, the system might merely inform you that there is a traffic jam ahead, and rather than turn left, which is your gut feeling, the system tells you to turn right. You know that the system has more information than you and so you trust it and turn right. At this point, you are still the one making the decision as to whether or not to accept the recommendation. The GPS system turns into your *agent* when you tell it where you want to go, and *the system decides* the best way to get there. You are no longer making that decision. At the stage of sovereign, the system realizes that, if it wants to avoid the traffic jam ahead, it cannot suggest to every driver the same alternative route because then the jam would be replicated on the alternative route. So the system makes the decision to inform half of the drivers of the alternative route while it keeps this information from the other half of the drivers. At that point the system has become sovereign. It proceeds by making independent decisions.

Examples could be multiplied, and Harari does present some more. But the message is already clear: "Once biologists concluded that organisms are algorithms, they dismantled the wall between the organic and the inorganic, turned the computer revolution from a purely mechanical affair into a biological cataclysm, and shifted authority from individual humans to networked algorithms."[52] Harari ends the chapter with the following warning: "If authority shifts altogether away from human beings into the hands of highly intelligent algorithms, then liberalism will collapse."[53] It will collapse because the authority to make decisions about our lives has been handed over to external algorithms that create—not necessarily an Orwellian future—but a centralized, networked system of "us" that will be run by those who run the computers. Harari calls this "new religion" Dataism. Dataism is only interested in human beings to the extent that they fulfill required functions. "By equating the human experience with data patterns, Dataism undermines our main source of authority and meaning."[54]

Outlook

The above scenarios portray two very different visions—and possibly fates—of modernity and humanism. For Feenberg, as becomes clear through his reinterpretation of Marcuse and Heidegger, the (modern) era of technological revealing is and ought to be focused on the protection of biological mankind and nature. Wherever these are threatened by purely means-oriented technical devices, the lifeworld with its "lessons of experience"[55] will step in by setting a limit (*peras*). This limit will influence the design of technical devices so as to guarantee that mankind can survive. Should this practice fail, democratic deliberation on design will add correctives. Feenberg's approach engenders an optimistic view of how mankind can keep technological progress in check. Harari fears that the advance of the new technologies and the new media means that modernity will potentially bleed into a post-human era that leaves the merely biological sphere behind. If the latest scientific discoveries turn out to be true, humanism, at least as we have known it, will come to an end. In the first scenario, mankind keeps authority over its experience and its world. In the second, it relinquishes experience, authority, and control.

Looking at these scenarios, both authors seem to agree on the singular logic to the process of civilization; namely, that mankind, through technological advance, increasingly externalizes and operationalizes the powers of the mind. If we look at its tools, it is hard to deny that mankind has undergone a fairly linear development from its early days to today, with the modern technological revealing, the *Ge-stell*, representing perhaps the tipping point of the Anthropocene. In describing this development, the term fate (*Schicksal*) seems less appropriate than Heidegger's term *Ge-schick*, a "sending" that has come over the earth. A *Ge-schick* is something that, although made by mankind, is not controlled by it, for it is hard to imagine that the technological revealing could be countered by mankind without incurring the greatest catastrophe. Technologies, whether exploitative or sustainable, are necessary for human survival. All this is part of what Heidegger meant by the *Ge-stell*, although he was hoping for a different mode of revealing that would replace its reign. The latest technologies seem to suggest that his hope will not be realized.

Feenberg, too, accepts this sense of the *Ge-stell* as *Ge-schick*, as evidenced by certain formulations (such as "the nature of our subordinate position in the technical systems that enroll us"),[56] or else in his conception of the level of primary instrumentalization in his critical theory of technology. But he rejects the autonomous nature of technology. Yet, in the sense described above, I would suggest that technological progress bears at least a semiautonomous, though nondeterministic character since we cannot go against the general direction of the Anthropocene without endangering human existence on the planet. But this interpretation does not necessarily imply dystopia. Or rather, it is dystopia only if we hypostatize a particular form of mankind's revealing as the benchmark for all time. Some of Heidegger's appeals to *Bodenständigkeit* (autochthony) point in this direction.[57]

In conclusion, I would like to return to the question as to what happens to "experience and subjectivity when they enter the age of technical media and the reality of a technological society and forms of life."[58] In his introduction to an essay collection entitled *Die technologische Bedingung* (*The Technological Condition*), Erich Hörl approaches, in the vocabulary of media theory, the central issue underpinning Harari's reflections. In his view, the contemporary technological condition, as manifested in the *Ge-stell*, is characterized by the fact that human agency is part of a more general state of affairs referred to as "environmental agency." The term environmental agency is supposed to capture the idea that agency is not reserved for human beings; rather, human agency is embedded in a bigger structure of "non-human agency distributed in the environment,"[59] which is provided, nourished, and maintained by media technologies. In the passage below, as throughout the section I am referring to, Hörl borrows the terminology of Mark B. N. Hansen, one of the contributors to the volume:

> Subjectivity is inherent to the sensory affordances of today's networks and environments, and in this way has become part of a generalized subjectivity, which also inheres in the data (*welche auch bereits Daten innewohnt*). "Literally clad in a multi-scaled and distributed sensory environment," writes Hansen,"our subjectivity of a higher order gains its power not because it perceives and orders that which is outside it but rather because of its immediate co-existence (*Mitteilhabe*) with or participation in the polyvalent agency of innumerable subjectivities."[60]

This view requires a fundamental reorientation of our sense of human agency, similar to the one we have seen in Harari's work since here human beings are no longer opposed as autonomous subjects to technical devices. Technical devices such as modern media connected to big data and networks spanning the globe gather, provide, and make digestible information that establishes a baseline of experiences and feelings upon which (in)dividual human experience builds. The all-important conclusion that the theoreticians of the technological condition draw from this assumption is this:

> Media no longer mediate our senses; rather they mediate our "sensory faculty itself" (*Empfindungsvermögen selbst*) which, augmented through digital designs, sensors, and smart chips, comes into being *before* any decisive separation into human and worldly sensory faculties takes place.[61]

This formulation claims—perhaps too emphatically—that our very ability to experience will, increasingly, first come into being on the background of media-based sensory input that precedes it. Such a development would, of course, spell radical change for human experience and human agency—if the media theorists are correct. Indeed, if it should come to pass that, prefigured by the "endlessly reproducible, all-pervasive *technical capacity of feeling* introduced by our "smart" tools and gadgets," we are no longer "the world's most complex sensing agent," then the human condition will truly have changed.[62]

Notes

1. Andrew Feenberg, *Between Reason and Experience*, Cambridge: MIT Press, 2010
2. Ibid., 220.
3. Ibid., chaps. 1 and 5.
4. "Converging Technologies for Improving Human performance." http://www.wtec.org/ConvergingTechnologies/Report/NBIC_report.pdf. Accessed December 2016.
5. Yuval Noah Hariri, *Homo Deus: A Brief History of Tomorrow*, London: Harvill Secker, 2016.
6. Feenberg, *Between Reason and Experience*, 195.
7. For the development of this side of the argument I am indebted, besides Harari's *Homo Deus*, to the short but meaty text by Norbert Bolz, *Das Gestell* (München: Fink, 2012).
8. Martin Heidegger, *The Question Concerning Technology and Other Essays*, Trans. and Intr. William Lovitt (New York: Harper & Row, 1977), 20.

9. Cf. the introductory paragraph of the essay "Technology and Meaning" in Andrew Feenberg, *Questioning Technology* (London: Routledge, 1999), 183
10. Feenberg, *Between Reason and Experience*, 181. This collection comprises nine essays written between 1992 and 2008 which were revised for this edition. The individual essays were published in diverse scientific, technical, and philosophical journals, in different countries, and for different occasions. The nature of the texts alternates between case studies, more common in the field of science and technology studies, and philosophical approaches to technology. The heterogeneity of these texts and the fact that they are spread out over time results, at times, in the repetition of themes and examples, but it also testifies to Feenberg's serious commitment to update and reformulate his ideas for a changing audience in an ever-changing vocabulary.
11. Feenberg, *Between Reason and Experience*, 165–66.
12. With this formulation we might, like so many commentators, get sucked into the vortex of Heidegger's endlessly self-mirroring terminology. It is to the great credit of Feenberg that he translates Heidegger's thought into a language of argument and reason that we can follow. I will try to do the same and only resort to Heidegger's own formulations when it is absolutely necessary.
13. Feenberg, *Between Reason and Experience*, 186.
14. Ibid., 190.
15. Ibid., 191.
16. Ibid., 193.
17. Ibid., 197. This quotation refers most directly to the notion of potentiality. Yet potentiality is precisely the concept that is needed to overcome the dystopian and deterministic aspects of the *Ge-stell*, and in that sense it is directly connected to the notion of essence elaborated here.
18. Ibid., 209.
19. See Heidegger, Martin, "The Question Concerning Technology," in *Martin Heidegger: The Basic Writings*, ed. David Farrell Krell (San Francisco: Harper, 1993[1963]), 29–31.
20. All quotations in this paragraph are from the essay "Impure Reason," in Feenberg, *Questioning Technology*, London: Routledge, 1999, here 201.
21. Ibid., 202.
22. Feenberg, *Questioning Technology*, 167.
23. Ibid., xiv.
24. Harari, *Homo Deus*, 56.
25. Ibid., 276–77
26. See footnote 4 above.
27. Ibid., 397.
28. Ibid., 85.

29. Ibid., 86.
30. Ibid., 327.
31. Ibid., 329.
32. Ibid., 306.
33. Ibid., 393. For a recent example, see: "Discrimination by Algorithm: Scientists Devise Test to Detect AI bias," *The Guardian Online*: https://www.theguardian.com/technology/2016/dec/19/discrimination-by-algorithm-scientists-devise-test-to-detect-ai-bias.
34. This is one of the points where I regret having to give short shrift to *Homo Deus*, for Harari also views humanism and liberalism as *religions* since, in his view, there never was any proof that humans were superior to animals for allegedly owning a soul or having free will. Both of these claims are given extensive discussion in his book.
35. Harari, *Homo Deus*, 352.
36. Ibid., 351.
37. Ibid., 389.
38. Feenberg, *Questioning Technology*, xiv.
39. Heidegger, "The Question Concerning Technology," 32.
40. Feenberg, *Between Reason and Experience*, xvii.
41. Ibid., 212.
42. Ibid., 215. Much of this argument on the reformulation of Marcuse's ideas depends on the use of the concept of *peras* (limit, boundary) in the revealing of essence in modernity. I am not sure that *peras* as limit/boundary in Heidegger's understanding can be equated, as Feenberg does, with the limit/boundary represented by the values of democratic deliberation in the creation of technical objects.
43. Harari, *Homo Deus*, 198.
44. Ibid., 237.
45. Ibid., 235.
46. Ibid., 276.
47. Ibid., 283–84.
48. Ibid., 285.
49. Ibid., 282.
50. Ibid., 305.
51. Ibid., 328–32.
52. Ibid., 345. Now it is true that Harari somewhat hedges his bets by stating that if the life sciences are wrong about all of this, i.e., "if organisms function in an inherently different way to algorithms, then computers [...] will not be able to understand us and direct our life, and they will certainly be unable to merge with us" (*Homo Deus*, 345). On the whole, however, he seems to be fairly convinced that they are right. As are many of the representatives of the transhumanist movement that he discusses in his last chapters.

53. Ibid., 350.
54. Ibid., 389.
55. Feenberg, *Between Reason and Experience*, xvii.
56. Cf. footnote 23.
57. Interestingly enough, however, Erich Hörl, in the introduction to *Die technologische Bedingung* (see footnote 58 below, here p. 7–8), discovers a different, forward-looking, sense to *Bodenständigkeit* in Heidegger's text on *Gelassenheit*. Here Heidegger differentiates between a *künftige* (future) and an *alte* (traditional) *Bodenständigkeit*, trusting that the *Bodenständigkeit* to come will reveal the as yet invisible meanings of the *Ge-stell*.
58. This quote, as well as the ideas presented here, is from the introduction to Erich Hörl (ed.), *Die technologische Bedingung. Beiträge zur Beschreibung der technischen Welt*, Frankfurt: Suhrkamp, 2011; here p. 50. All translations from the German are mine (AM).
59. Hörl, *Die technologische Bedingung*, 50.
60. Ibid., 51.
61. Ibid.
62. Ibid., 51–2.

Bibliography

Bolz, Norbert. *Das Gestell*, München: Fink, 2012.

'Converging Technologies for Improving Human performance." http://www.wtec.org/ConvergingTechnologies/Report/NBIC_report.pdf. Accessed December 2016.

Feenberg, Andrew. *Alternative Modernity. The technical Turn in Philosophy and Social Theory*, Berkeley: U of California Press, 1995.

———. *Between Reason and Experience*, Cambridge: MIT, 2010.

———. *Questioning Technology*, London: Routledge, 1999.

———. *The Philosophy of Praxis: Marx, Lukács, and the Frankfurt School*, London: Verso, 2014.

Harari, Yuval Noah. *Homo Deus. A Brief History of Tomorrow*, London: Harvill Secker, 2016.

Hörl, Erich (ed.), *Die technologische Bedingung. Beiträge zur Beschreibung der technischen Welt*, Frankfurt: Suhrkamp, 2011.

Heidegger, Martin. *Gelassenheit*, Pfullingen: Neske, 1959.

———. The Question Concerning Technology. In *Martin Heidegger: The Basic Writings*. Edited by David Farrell Krell. San Francisco: Harper, 1993[1963].

CHAPTER 11

Revisiting Critical Theory in the Twenty-First Century

Raphael Sassower

Introduction

We were optimistically mistaken to think that the work of the Frankfurt School and critical theory was accomplished by the end of the twentieth century. The flaws of the liberal-modernist state with its economic system of market-capitalism were laid bare by the time the global Great Recession of 2008–2012 was in full view. If the Habermasian attempt to reconcile old concerns with contemporary discursive options seemed reasonable—replacing the instrumental with the communicative rationality of the decades of prosperity in the Europe-American axis—by now we realize that more is needed to ensure a path toward any kind of radical change that offers human dignity or even economic (if not social and political) emancipation, personal or communal. Are the messianic dreams of emancipation still valid by the twenty-first century? How have these dreams been recalibrated in the American context in light of pragmatism and postmodernism?

R. Sassower (✉)
University of Colorado, Colorado Springs, CO, USA

D.P. Arnold and A. Michel (eds.), *Critical Theory and the Thought of Andrew Feenberg*, DOI 10.1007/978-3-319-57897-2_11

In what follows, Andrew Feenberg's contribution of critical theory in general and the philosophy of technology in particular are reviewed so as to shift the discussion to a postmodern fold. As will be seen, the classical immanent critiques of technoscience have to be modulated for the Digital Age, given the different material conditions of the time and the financial and institutional contexts within which they are debated. One must concede, though, that the original Marxist-laden terminology of the Frankfurt School all the way to its Americanized permutation in the hands of Herbert Marcuse and his student Andrew Feenberg are still relevant in the postcapitalist age.

The Philosophy of Praxis

The legacy of Marx was carried through in the twentieth century by what became known as the Frankfurt School. Feenberg is on solid ground when he describes the group that included Max Horkheimer, Theodor Adorno, Walter Benjamin, Herbert Marcuse, and Jürgen Habermas as being influenced by Georg Lukács and continuing the critique of modernity and the dominance of market capitalism. This group, he insists, turned away from "philosophical speculation toward action"—and this is in reference to the critical reception of what was perceived as Marx's overly (Hegelian) "messianic" or "utopian" streak—and remained mindful of finding an alternative to or being "in opposition to both Soviet 'dialectical materialism' and Western academic philosophy."[1] The Marxian "metacritique,"[2] one that "crosses the usual boundaries between philosophical and social explanation,"[3] has been transformed into an "immanent critique," one that still focuses on the classical problems of reification, alienation, exploitation, and the devastating consequences of rationality but which does not pretend to have a bird's eye view of reality or history.

As far as Feenberg's intellectual history goes, "The rational foundations of modernity are in crisis… under attack from two quarters, from fundamentalist religious ideology and from post-modern skepticism."[4] We will come back to these two trends in the concluding section when discussing the merits of postmodernism (and not any sort of fundamentalism) as fostering an emancipatory mindset. In the meantime, we should keep in mind what backdrop is assumed by the Marxist tradition: The Hegelian promise of the liberating powers of reason (and rationality) is false because its very own authority and restraining logical structure are themselves oppressive and dehumanizing.

Karl Marx of the early *Manuscripts* (1844) and of the later *Capital* deploys his philosophical analysis to undermine the pretenses of political economists, from Adam Smith and David Ricardo to others of his own time. But in doing so, Marx retains the influence and comprehensiveness of Kant and Hegel insofar that his own metacritique touches on historical developments, social formation, economic transactions, and the innovations provided by science and technology so as to bring them to bear on the specific historical developmental stage of capitalism. This tall order was reconfigured by the Frankfurt School without abandoning either its critical stance or its lofty goal of social transformation and personal emancipation. They had to account for Marx's recounting of Feuerbach's 11th thesis, "Philosophers have hitherto only *interpreted* the world in various ways; the point is to *change* it." With this in mind, and with the realities of Soviet communism since the Bolshevik Revolution of October 1917, they had to find an alternative that could still be considered Marxist through and through, but not devolve into a defense of the totalitarianism of State Socialism (Statism).

Feenberg connects Marx and Lukács and claims that "they consider philosophy to be the discipline in which the operative horizon of everyday life is raised to consciousness and subjected to rational criticism."[5] Here the commitment to rational critique (rather than romantic speculation) is clear, as well as the need to shift from conceptual abstractions to a personal level of self-consciousness where enlightenment is not global, but individual. He continues in the same paragraph to say that, "On this basis they argue that the conceptual dilemmas or 'antinomies' of philosophy are symptomatic of deep contradictions in social life."[6] Where Kant and Hegel were comfortable to analyze their cultural and natural surroundings conceptually, Marx and Lukács needed to ground these reflections in the realities of social life, the material conditions under which these conceptual contradictions (between subject and object, human and nature, thought and feelings, to name a few) were visibly manifested. The result of this grounding or anchoring of conceptual difficulties in the world of lived experiences "is the demand for a 'realization' of philosophy through the practical resolution of these contradictions."[7] For Feenberg, "This is perhaps the least understood aspect of the early philosophy of praxis of Marx and Lukács."[8]

With this in mind, Feenberg explains his own version of the "philosophy of praxis" as a logical process by which "abstractions" are "traced back to their roots in concrete social conditions" and those are eventually used to reconstruct their "revolutionary transformation."[9] In his critical assessment, Feenberg correctly identifies Adorno as more in the camp of utopian thinkers (continuing the German romantic and idealistic tradition), and contrasts him with his own teacher, Marcuse, who was eager to find a "new agent of revolution" in the New Left student movement of the 1960s in the US.[10] Marcuse was critical of the legacy he followed: "Marx promises a completely humanized nature but that project culminates in atomic bombs, not utopia. Lukács promises 'totality' in which objectivity is transparent to the social subject, but the outcome is totalitarianism."[11] And as Feenberg explains, "Critical Theory struggles against these utopian promises."[12] If utopian promises fail to deliver on their promises they could be forgiven because the material conditions of their time were not ripe enough, for example, or because human consciousness lagged behind the realities surrounding them; but when these utopian visions end up being destructive or totalitarian, then they ought to be discarded altogether. But can one offer practical changes without a blueprint? What would it mean for critical theory and the Frankfurt School to distance themselves from Marxist Communism?

In general, the main concerns of critical theory revolved around three axes: first among them is the philosophical concern with the construction of knowledge as objective (and empirically verifiable) as opposed to it being inherently and inevitably intersubjectively informed (socially and historically contingent). The rejection of a Hegelian absolute knowledge or its positivist counterpart between the two World Wars developed into an appreciation of the dialectical nature of the evolution of knowledge, and framing it requires different perspectives (instrumental or not). If knowledge is neither objective nor impartial, then its material grounding and its means-ends analysis must be a focal point of critical investigation. The Frankfurt School remained faithful to this philosophical concern, especially when its topics of study were cultural and political rather than scientific or economic.

The second axis was concerned with bringing theoretical insights into their practical spheres of relevance. This meant insisting on the starting point being social and economic, material, if you wish, rather than conceptual; it also meant that whatever logical or ideological "solution" could be offered, it had to be implemented in the real-world, among workers and consumers, viewers of the media and participants in the various political and labor institutions of the day. The fact that Hegel posited a perfectly harmonious congruence between reason and reality or that Marx could envision a reality where reason would protect humanity became an intellectual step that needed to be overcome by critical theory. If reason were to be regarded as important or worthy of privileging, it should do the heavy lifting of offering critical tools to dismantle institutions of oppression and put in their place the means by which emancipation becomes ever more possible. Faith in reason alone was not sufficient by the twentieth century; a more robust program had to be put in place (with or without an actual revolution of the appropriate class of alienated and exploited people).

The Marxist thrust of critically unmasking or exposing interests where they have been veiled behind the pretense of (scientific) abstraction—absolute categories and presumed value-neutral concepts—is revived under critical theory. At times this immanent critique is targeted to specific areas of culture, media, or communication (as seen in Horkheimer, Adorno, and Benjamin), at others it is more comprehensively carried out as a metacritique (as seen in the works of Habermas). But as the third axis reminds all who encounter texts written by the Frankfurt School, the thirst for freedom and emancipation is bound up with a revolutionary zest: one has to change consciousness so that the masses (however defined) will rise against their oppressors. After the Soviet experiment with its own revolutionary zeal, the European taste for a recurrent experiment was soured if not completely lost. As mentioned earlier, Marcuse could find some hope in the American New Left just as Habermas could appreciate the German student uprisings and the formation of the Green Party.

Feenberg is an astute and brilliant historian of these intellectual trends and movements associated with the Frankfurt School and their original source materials. For him, in addition to the axes listed above, one of the unifying themes has always been technology. According to him, "The Frankfurt school by contrast focuses on the consequences of the failure of resistance and the regressive forms of irrationality associated with it. The unifying theme of its analysis is the growing power of technology to control the social and natural worlds in an increasingly exploitive pattern signified by the concept of 'domination'."[13] Because of the fear of yet another utopian construct, they use "the logic of immanent critique and not a constructive alternative."[14] Their critique of the "technological consequences of the Enlightenment" is "not to blame technology for social ills, as does romantic critique, nor to appeal to technological rationality as an antidote to the inefficiency of capitalism, as does traditional Marxism, but to show how technology has been adapted *in its very structure* to an oppressive system."[15] Feenberg is correct to remind us that "it is impossible to return to what Marx once called the 'roasted pigeons of absolute science,' that is, to some sort of utopian or transcendental thinking."[16]

While some of the Frankfurt School directed their attention to cultural critiques, and while Habermas focused on communicative rationality as a means by which to ensure both the centrality of communicative rationality and a method by which consensus could be reached,[17] it is Marcuse who is most inspiring to Feenberg. To begin with, he could ascertain in the New Left a promising social agency with the "manifestations of solidarity and the rejection of consumerism."[18] Second, the insights of Marcuse's *One-Dimensional Man*[19] (1964) spoke to a whole generation of disillusioned thinkers who believed in the greatness of America post-World War II. Third, as Feenberg brilliantly explains, there is something profound in Marcuse's synthesis of the whole Hegelian Marxist tradition: bringing together "Lukács's concept of reification, Heidegger's critique of technology, Husserl's late discussion of science and the lifeworld, and Horkheimer and Adorno's theory of the impoverishment of experience under capitalism."[20] While the Marxist tradition insisted on the utopian potential of technology in its socialist manifestations to overcome the limitations of capitalism, "Marcuse insists instead that the existing technology dominates both human beings and nature."[21] As such, technology was oppressive and exploitive rather than liberating and humanizing.

And though he is probably one the most well-known disciples of Marcuse and definitely one of the most vocal defenders of the Frankfurt School, Feenberg clearly latches on this fourth, and perhaps the most relevant, axis of critical theory: technology. This is where Feenberg found his intellectual home, where he has made a contribution that transcends Marxian utopianism about the power of technology to overcome all the ills of capitalism, Heideggerian romanticism and vagueness, as well as the rest of the Frankfurt School's lamentation that is laced with messianic traces[22] he continues single-handedly to further the Marcusian critique of the technological dependence of humanity and the false promises it has in store for generations to come. Feenberg believes that "science and technology would be appropriated in the context of the socialist transformation," but explains that "to save Enlightenment from itself they must overcome the damage capitalism has inflicted on experience and reason."[23] Two issues are intertwined in these statements: first, the Enlightenment Project, as Habermas phrases it, has not been completed yet, and it is the responsibility of the extended family of the Frankfurt School to bring it to fruition; and second, just because science and technology—what some call technoscience—have been co-opted under capitalism to devastating effects does not mean that they cannot be used for good as well, that is, to bring about the completion of the Enlightenment Project under the material and social conditions of socialism.

Outside the small cottage industry that has evolved around Habermas across the Atlantic divide and beyond, Feenberg remains quite alone in the Americas as the promoter of what is left of the Frankfurt School, especially his devotion to his teacher, Marcuse. Stripping away the nostalgia for Heidegger's antitechnology romanticism (and its accompanying dangerous flirtation with fascism and outright anti-Semitism), Feenberg is at his best when focusing on technology. His *Critical Theory of Technology* of 1991 established him as an active participant and founding voice of what we call the philosophy of technology. With a keen Marxist eye to contemporary hubris about environmental crises and the onslaught of the Digital Age, he remains a critical analyst committed to the Marxian hope for distilling the power of technology to bring about human emancipation. If Marx was too utopian and the Frankfurt School too dystopian, Feenberg wants to restore our faith that under democratic conditions technology can serve humanity well: "technology does not pose an insuperable obstacle to the pursuit of 'humanistic' values. There is no reason why it could not be reconstructed to conform to the needs of a socialist society."[24]

Feenberg offers his "critical theory of technology" as an alternative to the two dominant theories, the "instrumental" one whereby technology is a tool for furthering the values expressed in political and cultural spheres, and the "substantive" one which claims a certain level of autonomy to technology as setting its own values or practices in motion regardless of what aims are set for it.[25] On his view, "technology is not a destiny but a scene of struggle. It is a social battlefield, or perhaps a better metaphor would be a *parliament of things* on which civilizational alternatives are debated and decided."[26] In this so-called "parliament of things" presumably debates will ensue in ways that could allow a consensus to come about, and if not, a simple majority would rule the day, as democratic principles will be in place and the rule of law would defend their implementation.

Since Feenberg's critical theory of technology is part and parcel of what he calls the philosophy of praxis, and since he believes that, "the philosophy of praxis avoids the crude economic reductionism of which Marxism is often accused by reconstructing the relations idealism established between its categories as social relations,"[27] one wonders what conceptual apparatus he provides for capitalism and socialism. For him, they "are rather ideal-types lying at the extremes of a continuum of changes in the technical codes of advanced societies"[28] and therefore not on an evolutionary or developmental trajectory in the classical Marxist sense. With this in mind, let us shift the brief exposition of Feenberg's views to two specific areas: technoscience and political economy.

Technoscience and Political Economy

Though there are quite a few of us who have mindfully used the term technoscience (without hyphen) for some time, there are those, like Feenberg, who find this usage to be offensive on the ground that these two categories, science and technology, operate in fact with two different kinds of logic, the former more holistic in a Hegelian sense and the latter more "purposeful" or "instrumental" in the Weberian sense of instrumental rationality.[29] Since Feenberg is interested in "political reform for technology and reform from within for science," it stands to reason that he asks right away the following question: "if techno-science is a single phenomenon, on what basis can one make this strategic distinction?"[30] He then attributes to Habermas and Marcuse two related views he finds mistaken: "In fact, critical theory tends to waver uncomfortably between a utopian politics of techno-science (Marcuse), and acceptance of the neutrality of techno-science in its proper sphere (Habermas)."[31] Without a distinction between science and technology, he concludes "we will be unable to put forward a believable case for a critique and transformation of modern forms of rationality."[32]

To begin with, "technoscience" is neither a simple convenience that incorporates both science and technology nor a shorthand designation that overlooks some differences between science and technology. On the contrary, the label technoscience has been in vogue because it encapsulates several historically informed insights: First, there is no chronological trajectory between the one and the other such that, as the standard view claims, science is theoretical (or pure) and technology its eventual application or handmaiden. Instead, the one informs the other, the one precedes the other in nondeterministic ways so that the priority accorded the one over the other loses its potency. The reciprocity that entangles the one with the other blurs any clear-cut lines of demarcation or delineation that are conveniently (but erroneously) drawn between the two.[33]

Second, since there is no chronological priority of science over technology, and since the participants in both areas find it difficult at times to distinguish their activities as being scientific rather than technological (one can cite here the Manhattan Project just as much as the Human Genome Project and the activities at CERN), one wonders what is at stake in adhering to a line of demarcation between science and technology. The standard view has claimed that not only different logics are evident (in Feenberg's sense), but, more importantly, while we can ascribe moral culpability to technology—from the use of Zyklon B to annihilate Jews during the Holocaust to the dropping of atomic bombs on Hiroshima and Nagasaki (Are military leaders at fault? Are politicians?)—theoreticians can be insulated from such responsibility. The critique of this useful distinction informs the insistence of stating that what we are dealing with is technoscience, an amalgam of participants in the scientific community whose activities ought to be questioned prior to any project so as to ensure that unintended consequences are fully considered before engineers and technicians are put in place to fabricate bridges or bombs.

Third, if critical theory as immanent critique of both science or technology is to be practiced, it would readily submit that lines of demarcation are so blurred that holding onto them is a conceptual fiction, an idealization reminiscent more of Weberian analyses than Marxian. This would mean, for example, that once our attention was drawn to the industrial–academic–military complex we would find it difficult, if not impossible, to label this person a scientist while that one a technologist (technician or engineer). Does the designation depend on one's academic credentials? Or does it rather depend on the source of funding? Does it depend on whether or not a patent is filed for or whether an implementation is guaranteed? These are rhetorical questions, but they highlight the untenable position the standard view of science and technology attempts to defend. The material conditions of the practices and activities that come under the rubric of technoscience are complex, fluid, and for that reason employ a variety of logical constructs and social formations (for profit, nonprofit, public, and private).

Given these three facets of technoscience it is reasonable to lump scientific and technological activities into technoscience and ensure that this broader term or label comes under full critical scrutiny that could be avoided by focusing partially on either of its components on their own. As I have studied technoscientific practice over the decades a few observations have come to light. Some of these would shed a critical light on Feenberg's own views while others would be complementary, filling some gaps left by his more Hegelian-Lukácsian interpretation of Marx. I should submit upfront that my own appreciation of technoscience is informed by Marxism, Popperianism, and postmodernism (admittedly not common conceptual and intellectual bedfellows). In what follows, I suggest to retrieve the most salient elements of a critical (rational) approach to technoscience, without thereby limiting its scope.

If classical Marxism envisioned technoscience as the means by which social formations in a postcapitalist world could be rearranged to ensure greater liberty and equality, if it also appreciated technoscience as the engine that would ensure the surpluses of capitalist production and transform it into socialist distribution and consumption, then it was technophile in its orientation and outlook. By contrast, the Frankfurt School crowd turned this technophilia into technophobia, claiming that no good could come of an over-scientific mode of production, reproduction, distribution, and consumption. The divide between technophobes and technophiles is as long as technoscience is old, and in this sense critical theory replays a familiar game between those who are wary of anything technoscientific and those who endorse any of its manifestations.

But lining up all the examples and arguments for and against technoscience remains an abstract exercise despite its empirical grounding in case studies. As much as statistical data (about climate change or the efficacy of this or that drug treatment) can shock the public and even put pressure on politicians to change public policy, it is prone to manipulation and the corrupting effects of framing, as behavioral economists remind us.[34] And here classical Marxism (and not its Hegelian antecedents or Lukácsian interpretation) comes into play: if we are to examine technoscience we ought to appreciate its political economic framing, its actual institutional and financial apparatus. (In this context, leftist sociologists from C. Wright Mills to Stanley Aronowitz have made significant contributions outside the label of the Frankfurt School or critical theory). To speak of the underlying rationality of technoscience is philosophically barren not because it is irrational, but instead because its rationality is context specific: its externalities affect its internal workings such that a national exigence, such as the impending threat of fascism in World War II or the Cold War between the US and its allies and the Soviet Union and its allies, transforms the very logic once claimed for it. This entails the rethinking of what is under scrutiny: technoscience or the political economy within which it is practiced.

If the material conditions of capitalism are examined, what we find is that classical market capitalism is quite different from its current operation. Though Marx's critique of Adam Smith's classical capitalism is outdated in many ways, some of its principled arguments still hold true, as Eli Zaretsky[35] and David Harvey[36] remind us. There are endemic contradictions in the system, from monopolistic tendencies to ongoing cyclical crises (the most recent of which was the Great Recession of 2008–2012). Technoscience is part and parcel of the ongoing transformation of capitalism such that its very operation depends on changing funding formation, from private industry to government-sponsored research and development for military or other uses. How different institutions interact in this maze of activity remains open to interpretation, as the economist Philip Mirowski[37] astutely argues, and therefore grand statements about "capitalism" or "technoscience" are barren and misleading. What is at issue at this juncture is not some philosophical understanding of technoscience—metacritique or immanent, but a nuanced study of particular projects and groups and the specific institutional and funding conditions under which technoscience operates in the real world of the industrial–military–academic complex.

Shifting the study of technoscience to its political-economic domains is crucial because it also speaks to the kind of image or vision the Frankfurt School upholds in regards to the potential consensus that may be reached among citizens in a democracy (Habermas) or the conditions for social transformation (Marcuse and Feenberg). It is in this context that Karl Popper's debates with Marcuse,[38] for example, come to light. What is at stake here is the question of holism versus piecemeal engineering. Popper accuses the Frankfurt School of offering a holistic vision that requires a revolution, which is for him methodologically flawed (because only individual actors ought to be considered in formulating policies) and politically dangerous (because of totalitarian tendencies and the violence that accompanies such revolutions). In their exchange, Marcuse talks about the "integration of the working class" that eventually undermines the revolutionary zest of the Marxian sort.[39] For him, a revolution ferments "out of disgust at the brutality and ignorance of human beings."[40] But this fermentation of discontent does not necessarily mean that a revolution is about to erupt. Marcuse is careful to argue that even with minor reforms, "sooner or later the point is reached where reforms run up against the limits of the system: where to put through reforms would be to sever the roots of capitalist production—namely profit."[41]

What becomes clear from these debates is that the disagreement is not about goals—they both wish for Enlightenment-like emancipation—but rather the means by which to accomplish these. Popper's response to Marcuse is that indeed "the state is a necessary evil; without a state, things won't work."[42] His piecemeal engineering, his trial and error method as applied to political matters, is informed by his methodology of science (conjectures and refutation). But he is quick to assert: "While positivism teaches: 'Stay with what can be perceived,' I teach: 'Be bold in erecting speculative hypotheses but then criticize and examine them mercilessly'."[43] Revolutions promise too much, deliver too little, and in their process are not only violent but may also find themselves at a place different from the one originally envisioned. The point of this exchange between Popper and Marcuse is not only to highlight their differences, but also to keep in mind that they are not talking at cross-purposes, but were both affected by the traumas of the two World Wars in Europe. It was their philosophical duty to address matters of life and death, namely, the centrality of technoscience in forming political institutions that would protect humanity and ensure world peace. Whose prescription ends up to be more faithful to actual political processes remains open, since Marcuse is absolutely correct that all the reforms in the world eventually have to culminate in changing the basic framework of a system (such as replacing profit maximization by sustainability). And here, once again, it is Habermas of late[44] whose critique of "technocracy" has more currency than anything that still comes out of the disciples of Marcuse, because he fully accounts for the European Union's economic trappings within a political and legal framework that fails to be democratic no matter how "technocratically" efficient it appears.

There are few scholars nowadays who are comfortable (both in terms of background and in terms of temperament) to enlist the insights of those from across disciplinary and methodological divides, and Feenberg must be counted as one of the best of them. The Analytic-Continental divide in philosophy departments set the stage for further divisions between other schools of thought. The divisions are not simply in light of referencing one school of thought rather than another, but more so in being able to apply the insights of one group to the concerns of the other, as Feenberg has done,[45] for example, in his treatment of modernity and the social sciences, and as such is quite an exception to the silo-like outlook of others. These silos of knowledge ensure that cross-fertilization rarely takes place and that their own echo chambers support lively but sterile internal conversations. If Popper and Marcuse exchanged ideas, their respective students and disciples neither talk to each other nor cite each other's works. If we move away from philosophical abstractions and examine the economic conditions under which technoscience actually produces and distributes the fruits of its research and development, the Heideggers of yesteryear should recede into oblivion because they insist on a jargon of obstructionism that defies exchanges with sociologists of knowledge, for example, or those studying the social framing of science and technology. And if there is a lamentation to be explored, it is not of the loss of the emancipatory (utopian) power of technoscience, but rather one that explains the backdrop of ideals against which technoscientific practices are to be ideally and concretely measured.[46]

If contemporary capitalism no longer resembles the contours of classical capitalism, and if technoscience is not simply an engine for economic growth and progress, then we have two interrelated practices and institutions that reciprocally transform each other. Just like the problems of measuring quantum states without freezing in time and space some features of the particles under scrutiny (Heisenberg's Uncertainty Principle), so is the case with postcapitalist technoscience: What variables and features are to be suspended in time and space in order to measure others? Under what conditions would one claim progress rather than regress? What values are being promoted or discarded when examining the current state of affairs: population growth, disease prevention, income per capita, or wealth and income inequalities? Rationality as such cannot even begin to answer any of these questions; and without answering these questions, the sterility of an immanent critique of technoscience (that does not account for the economy) becomes clear. In other words, any analysis of technology without considering its technoscience remains wanting; and any examination of technoscience without its economic context is wanting as well. Only with this in mind can we hope to find political solutions to our problems, public policies that promote human integrity and freedom, personally and socially.

Before moving to the next section, I must also acknowledge that the political domain by now is financially dependent on the largesse of economic interests (the elite 1% as we call them now) whose objective is the retention of the consolidated power-relations between the wealthy few and the political and legal protective shields ensured by elected politicians. So, to speak of politics in the US in the twenty-first century, as Lawrence Lessig[47] brilliantly summarizes the current corruption of American politics, is to speak of the economic influence of wealthy individuals and corporations and the disproportionate and undemocratic power they have amassed. And here Marx and Popper seem to be on the same side, for they both argue for the freedom of the population to make decisions in an open society wherein everyone can and will pursue whatever interests are on their mind, a utopian dream that should be achieved. As I have argued elsewhere,[48] Popper's political views are not liberal in the narrow sense of the Austrian School, but much more fluid in their application. Just as Marx would not endorse the totalitarian regime of the Soviet Union so Popper would not endorse the neoconservative policies of American Republicans. They both appreciated a basic human decency and dignity that the State ought to promote and protect.

Postmodern Voices

Despite its maligned reputation by the end of the twentieth century as being too vague, unoriginal, and even politically irresponsible, postmodern practices have been fruitful and informative for our present discussion.[49] It was Jean-Francois Lyotard who took to task technoscience as a self-legitimating discourse, one among many.[50] The collusion between capitalism and technoscience was explained in terms of power relations and the ways in which one made room for the other and eventually ensured the concentration of power and domination. It makes sense in a discussion of Feenberg to recap and develop the postmodern tension that is introduced to critical theory in current debates. Though Feenberg mentions the "postmodern" condition in terms of a "post-Marxist" sense,[51] we may want to pursue a more critical appreciation of the postmodern approach to the critical analysis of social and economic conditions, one that transcends class issues and the conditions of social realities. One may recognize the current discussion not as a difference in kind but simply as a difference in the degree with which one is willing to pursue the postmodern agenda more fully, taking it to some of its logical consequences beyond the critical theory matrix. Similar to the agenda of the critical theorists, postmodern critics wanted to uncover not only the rationality on which technoscience garnered its capitalist support but also the discursive moves both made in response to each other's needs and prospects. Though French speaking, for the most part, postmodern critics studied the same texts as their German-speaking counterparts, and they all grew disillusioned by the horrors of Soviet Communism and the specter of two World Wars on their soil. But common heritage, both intellectual and geographical, was not the only thing they shared.

They also shared a sense of responsibility to improve the fate of humanity. If philosophy could be of use, so much the better; if it came short, it should be discarded. With this mindset, postmodern critics had no problem giving up on the Enlightenment Project that the Frankfurt School was still trying to complete (at least Habermas has been vocal about it in Europe and Feenberg in the US). As far as they were concerned, modernity could be "displaced" by postmodernity; absolute principles and foundations would be displaced by a plurality of temporal contexts; and rationality as such would be supplemented by the notion of reasonableness. Displacement, to be sure, is different from "replacement" insofar as the former allows for the simultaneous applicability of different principles given different conditions rather than the latter's requirement that the one is fully discarded and replaced by a better, improved, or more fully enlightened one. In other words, instead of demanding "either/or," there is an allowance for "both/and." This may sound like irresponsible gibberish that verges on radical relativism, the kind made famous by Paul Feyerabend[52] in his "anything goes" version of the methodology of science. But this is not the case.

On the contrary, instead of being relativistic and irresponsible, postmodern critics must explain the contexts under which they offer their critical assessments and remain more modest in offering solutions because they recognize the limitations of the contexts from which they are drawn. As mentioned earlier, Heisenberg's Uncertainty Principle suggests that it is impossible to measure both the position and velocity of particles in quantum states; likewise, it is impossible to provide a full technoscientific picture of socioeconomic realities and the individuals who experience them. To say anything critical and informative, one must lower the expectations and refine the assessment similar to the case-by-case basis used in courts of law.[53] This way of postmodern practice also avoids the reductionism ascribed to Marxist metacritiques by critics and scholars like Feenberg. Postmodernist critiques then have the advantage of a plurality of perspectives all of which should contribute to a fuller picture of the situation with modest expectations but great emancipatory hopes.

It was Popper who originally explained the need for the logic of a situation in order to assess its contours and the solutions appropriate to that epistemological matrix (later taken up by feminist critics, such as Donna Haraway, in their situational epistemology).[54] And it was Marx long before him who insisted that whatever one claims as knowledge must be socially and economically situated so as to reveal the interests of those propounding this or that theoretical view. Postmodernists then combine these strong claims not in order to rescue some notion of rationality or reason, or of liberty and equality, but instead in order to appreciate their historical contingencies and the inherent impossibility of permanent and final judgments, especially as they pertain to the social and human spheres.

Postmodern critics just like critical theorists focus on technoscience in its various manifestations, from research laboratories to media outlets, always cognizant of their diversity. This means, for example, that the Marxist critique of the Industrial Revolution had quite different technoscientific practices to contend with than those of the Digital Age, as Yochai Benkler,[55] for one, reminds us, and therefore the toolbox of classical Marxism and early Frankfurt School may only partially be of use today. Given the different contexts and realities of technoscience, it is unreasonable to argue about climate change in the ways manufacturing was argued about a century ago. Maybe not a completely new toolbox is needed, but at least different tools within that toolbox, the ones that can more readily be applied in the present context of the Internet and digital technologies, of outsourcing and crowdfunding.[56] This is not to say that certain classical Marxian principles cannot be deployed now as they were then, such as the notion of alienation and exploitation and the unfairness of income and wealth inequality. But the binary of capitalists/proletariat may have to change in the complex social stratification we encounter today. Yes, the "1% vs. 99%" is still a conceptual image that expresses correctly the frustration over income and wealth inequalities in the US; but is it as informative as the binary of capitalist/proletariat was in the middle of the nineteenth century? What does the "1%" represent in contemporary parlance? This elite of super-wealthy billionaires not only controls industry and finance alone, but is also social insulated from the travails of the rest of us and exerts a certain political and ideological control only partially exercised in the Gilded Age of the late nineteenth century.

Though Marcuse's critique of Popper is formidable,[57] what is missing from this critique and what postmodernism brings to the fore is the realization that the debate over "holism" versus "methodological individualism" is really about Meta-narratives (or metanarratives) and narratives. That is, critical thinkers may have appreciated the nuances of the social sciences methodologically speaking, but they have not yet relinquished a "Meta"-critique, one that knows it all, that is cocksure about its contours and future, and one under whose theoretical umbrella everything must neatly fit. The postmodern critics unwittingly have understood the Popperian antipathy to grand-narratives and have responded in turn. Their studies are more circumspect; their analyses more varied; and their understanding of the latest burst of the housing bubble and the Great Recession that ensued is not limited to one explanation, because too many variables contributed to this disaster. Offering at this stage a Grand-Marxist-Capital-vs-Proletariat-narrative would do too much and too little: it would promise a definitive answer to a complex puzzle and it would thereby overlook too many contributing factors that allowed for this multifactorial phenomenon to deprive millions of their homes and jobs. A "smoking gun" or "silver bullet" would be a convenient Marxist or Frankfurtian end point; instead, we must contend with the meandering, never fully complete postmodern set of narratives to thread this quilt together.

There are some excellent examples of how a Foucauldian analysis is practiced without even mentioning Foucault or deconstruction or discursive analysis. For example, Steve Fraser recently illustrated the difference in the responses to the excesses of the two Gilded Ages, the first of the late nineteenth century and running into the twentieth century, and the second that began in the Reagan era and lasted till the financial collapse of 2008, in linguistic terms as much as in sociological ones that account for the condition and power of the labor movements during those periods. He specifically speaks of "linguistic cleansing" that has taken place by the twenty-first century so that "class warfare" or any other Marxist terms could not be mentioned in polite company[58]; likewise, he explains how "employees" who could command wages and benefits (and legal rights) were transformed into "independent contractors" without any benefits and with no regulatory protection, all in the name of "entrepreneurship" and individual choice and freedom.[59] If classical Marxism and the Frankfurt School and critical theorists still hoped for the self-determining labor "class" to fight its ideological wars (and to succeed in bringing about a revolution), by the time we think about the discursive and experiential realities of the twenty-first century, old linguistic and conceptual canons look rusty if not inoperative. And the ideological twists of capitalism and socialism are no longer "ideal-types" (as Feenberg sees them), but outdated tropes whose historical value remains relevant but not as much as it was a century ago.

Those of us who work under the label of postmodern technoscience appreciate the Marxian dictates to focus on material conditions, on the realities and experiences of people who actually work for a living and who have to contend with the limitations of the State; likewise, we appreciate the concern for the privileged position of technoscience and its discursive power into the present. If any kind of Popperian piecemeal engineering is possible, if we attempt to improve the human condition with the knowledge we have accumulated over the past few centuries, we ought to retain our critical posture and neither be seduced by unattainable ideas nor by meager accomplishments that change very little. We must remain open-minded and mindful of alternative proposals and ways of seeing the world. Believing in our own infallibility is false and irresponsible. In the spirit of modesty, postmodern critics offer their analyses with an expectation that their judgments will be improved upon, even displaced. But as long as they are not replaced once and for all, their eventual revival may offer new insights long forgotten. And as they come about, one incomplete narrative at a time, they allow all of us to become participants in the construction of our destiny. It is in this spirit that Feenberg's contributions still have much to teach us.

Notes

1. Andrew Feenberg, *The Philosophy of Praxis: Marx, Lukács and the Frankfurt School* (London and New York: Verso, 2014), 151.
2. Ibid., 9.
3. Ibid., 11.

4. Ibid., vii.
5. Ibid., xiii.
6. Ibid.
7. Ibid.
8. Ibid.
9. Ibid., xiv.
10. Ibid.
11. Ibid., 152.
12. Ibid.
13. Ibid., 153.
14. Ibid., 155.
15. Ibid., 164.
16. Ibid., 171.
17. Jürgen Habermas, *The Theory of Communicative Action*. Translated by Thomas McCarthy (Boston: Beacon Press, 1981).
18. Andrew Feenberg, *The Philosophy of Praxis: Marx, Lukács and the Frankfurt School*. (London and New York: Verso, 2014), 173.
19. Herbert Marcuse, *One-Dimensional Man: Studies in the Ideology of Advanced Industrial Society* (Boston: Beacon Press, 1964).
20. Andrew Feenberg, *The Philosophy of Praxis: Marx, Lukács and the Frankfurt School*. (London and New York: Verso, 2014), 175.
21. Ibid., 178.
22. Ibid., 194.
23. Ibid., 195–6.
24. Andrew Feenberg, *Critical Theory of Technology*. (New York: Oxford University Press, 1991), 127.
25. Ibid., 5.
26. Ibid., 14.
27. Andrew Feenberg, *The Philosophy of Praxis: Marx, Lukács and the Frankfurt School* (London and New York: Verso, 2014), 204.
28. Andrew Feenberg, *Critical Theory of Technology* (New York: Oxford University Press, 1991), 89.
29. Ibid., Chap. 8.
30. Ibid., 165.
31. Ibid.
32. Ibid.
33. Raphael Sassower, "Technoscience and Society," *Encyclopedia of the Philosophy and the Social Sciences*, edited by Byron Kladis (New York: Sage Publications, Inc., 2013).
34. Richard H. Thaler, *The Winner's Curse: Paradoxes and Anomalies of Economic Life* (Princeton, NJ: Princeton University Press, 1992). Daniel

Kahneman, *Thinking, Fast and Slow* (New York: Farrar, Straus, and Giroux, 2012).
35. Eli Zaretsky, *Why America Needs a Left: A Historical Argument* (Cambridge, UK: Polity Press, 2012).
36. David Harvey, *Seventeen Contradictions and the End of Capitalism* (New York: Oxford University Press, 2014).
37. Philip Mirowski (2011), *Science-Mart: Privatizing American Science* (Cambridge and London: Harvard University Press).
38. A.T. Ferguson, ed., *Revolution or Reform? A Confrontation* (Chicago: New University Press, 1976/1972).
39. Ibid., 66.
40. Ibid., 72–3.
41. Ibid., 74.
42. Ibid., 87.
43. Ibid., 98.
44. Jürgen Habermas, *The Lure of Technocracy.* Translated by Ciaran Cronin (London: Polity, 2015/2013).
45. Andrew Feenberg, *Alternative Modernity: The Technical Turn in Philosophy and Social Theory.* (Berkeley: University of California Press, 1995); see also, Andrew Feenberg and Alastair Hannay, eds., *Technology and the Politics of Knowledge.* (Bloomington and Indianapolis: Indiana University Press, 1995).
46. Raphael Sassower, *Compromising the Ideals of Science* (Hampshire, UK: Palgrave Macmillan, 2015).
47. Lawrence Lessig, *Republic, Lost: How Money Corrupts Congress—and How to Stop It.* (New York: Twelve, 2011).
48. Raphael Sassower, *Popper's Legacy: Rethinking Politics, Economics, and Science* (London: Acumen Publishers and Montreal: McGill-Queen's University Press, 2006).
49. Ingeborg Hoesterey, ed., *Zeitgeist in Babel: The Postmodernist Controversy.* (Indianapolis: Indiana University Press, 1991).
50. Jean-François Lyotard, *The Postmodern Condition: A Report on Knowledge.* Translated by Geoff Bennington and Brian Massumi (Minneapolis: University of Minnesota Press 1984/1979).
51. Andrew Feenberg, *Critical Theory of Technology.* (New York: Oxford University Press, 1991).
52. Paul Feyerabend, *Against Method.* (New York: Verso Books, 1975).
53. Jean-François Lyotard and Jean-Loup Thébaud, *Just Gaming.* Translated by Wlad Godzich (Minneapolis: University of Minnesota Press, 1985).
54. Raphael Sassower, *Popper's Legacy: Rethinking Politics, Economics, and Science* (London: Acumen Publishers and Montreal: McGill-Queen's University Press, 2006), Chap. 3.

55. Yochai Benkler, *The Wealth of Networks: How Social Production Transforms Markets and Freedom* (New Haven and London: Yale University Press, 2006).
56. Raphael Sassower, *Digital Exposure: Postmodern Postcapitalism* (Hampshire, UK: Palgrave Macmillan, 2013).
57. Herbert Marcuse, "Karl Popper and the Problem of Historical Laws," in *From Luther to Popper* (1972). Translated by Joris De Bres (London and New York: Verso, 1975), 191–208.
58. Steve Fraser, *The Age of Acquiescence: The Life and Death of American Resistance to Organized Wealth and Power* (New York: Little, Brown and Company, 2015), 6.
59. Ibid., 347–357.

Bibliography

Benkler, Yochai. *The Wealth of Networks: How Social Production Transforms Markets and Freedom*. New Haven: Yale University Press, 2006.

Feenberg, Andrew. *Alternative Modernity: The Technical Turn in Philosophy and Social Theory*. Berkeley: University of California Press, 1995.

———. *Critical Theory of Technology*. New York: Oxford University Press, 1991.

———. *The Philosophy of Praxis: Marx, Lukács and the Frankfurt School*. New York: Verso, 2014.

Andrew Feenberg and Alastair Hannay, eds. *Technology and the Politics of Knowledge*. Bloomington: Indiana University Press, 1995.

Fraser, Steve. *The Age of Acquiescence: The Life and Death of American Resistance to Organized Wealth and Power*. New York: Little, Brown and Company, 2015.

Habermas, Jürgen. *The Theory of Communicative Action*. Translated by Thomas McCarthy. Boston: Beacon Press, 1981.

———. *The Lure of Technocracy*. Translated by Ciaran Cronin. London: Polity, 2015/2013.

Harvey, David. *Seventeen Contradictions and the End of Capitalism*. New York: Oxford University Press, 2014.

Hoesterey, Ingeborg, ed. *Zeitgeist in Babel: The Postmodernist Controversy*. Indianapolis: Indiana University Press, 1991.

Kahneman, Daniel. *Thinking, Fast and Slow*. New York: Farrar, Straus, and Giroux, 2012.

Lyotard, Jean-François and Jean-Loup Thébaud, *Just Gaming*. Translated by Wlad Godzich. Minneapolis: University of Minnesota Press, 1985.

Lessig, Lawrence. *Republic, Lost: How Money Corrupts Congress—and How to Stop It*. New York: Twelve, 2011.

Marcuse, Herbert. "Karl Popper and the Problem of Historical Laws." In *From Luther to Popper* (1972). 191–208. Translated by Joris De Bres. London and New York: Verso, 1975.
———. *One-Dimensional Man: Studies in the Ideology of Advanced Industrial Society* Boston: Beacon Press, 1964.
Sassower, Raphael. *Compromising the Ideals of Science*. Hampshire, UK: Palgrave Macmillan, 2015.
———. *Digital Exposure: Postmodern Postcapitalism*. Hampshire, UK: Palgrave Macmillan, 2013.
———. "Technoscience and Society." *Encyclopedia of the Philosophy and the Social Sciences*. Edited by Byron Kladis. New York: Sage Publications, Inc., 2013.
———. *Popper's Legacy: Rethinking Politics, Economics, and Science*. London: Acumen Publishers and Montreal: McGill-Queen's University Press, 2006.
Thaler, Richard H. *The Winner's Curse: Paradoxes and Anomalies of Economic Life*. Princeton, NJ: Princeton University Press, 1992.
Zaretsky, Eli. *Why America Needs a Left: A Historical Argument*. Cambridge, UK: Polity Press, 2012.

CHAPTER 12

Andrew Feenberg, Critical Theory, and the Critique of Technology

Douglas Kellner

From the 1990s through the present, Andrew Feenberg has systematically developed a critical theory of technology that charts the development and increasing importance of technology in all facets of our lives and that develops a critical theory that articulates how technologies can serve as instruments of domination or emancipation and democratization. This dialectical optic concerning technology and the contemporary world is grounded in the work of the Frankfurt School and Feenberg's teacher Herbert Marcuse.

In *Critical Theory of Technology* (1991), Feenberg draws on neo-Marxian and other critical theories of technology, especially the Frankfurt School, to criticize determinist and essentialist theories. In this groundbreaking work (which went into its second edition in 2002), he discusses both how the labor process, science, and technology are constituted as forms of domination of nature and human beings, and how they could be democratically transformed as part of a program of radical social transformation.

D. Kellner (✉)
UCLA, Los Angeles, USA

D.P. Arnold and A. Michel (eds.), *Critical Theory and the Thought of Andrew Feenberg*, DOI 10.1007/978-3-319-57897-2_12

In *Alternative Modernity*,[1] Feenberg turns to focus on constructivist theories and the ways in which individuals and groups can reconstruct technology to make it serve more humane and democratic goals. This text draws on his earlier work while polemically developing his own positions within contemporary debates over technology. Feenberg's work dramatizes the importance and impact of technology within the contemporary world and the need to critically embrace and reconstruct technology as part of creating a more democratic, ecological, and humane society. His studies undertake to explore the impact of technology on diverse regions of human life and culture, and to interrogate both major theories concerning technology and some other cultural responses to the development of Western technology, including science fiction, dystopic film, and Japanese culture.

In particular, Feenberg uses a reconstructed version of the Frankfurt school critique of technology, building on his own earlier *Critical Theory of Technology*.[2] Yet Feenberg also draws on French postmodern theory, as well as Japanese theory and various cultural texts, to analyze Western modernity and to explore multicultural alternative modernities. His goal is to counter dystopic and technophobic visions of modernity, while showing some positive uses of technology to advance human emancipation and some alternative attitudes to and conceptions of technology and Western modernity.

Feenberg's major focus and distinctive position within current debates on technology emphasizes the democratic potential for the social reconstruction of technology. Feenberg rejects both neutralist positions which see technology as a mere instrument of human practice, amenable to any and all projects and uses, and determinist notions, which see it as a instrument of domination in the hands of ruling elites whose very construction determines the uses, limits, and applications of technology. Instead he sees technology as a contested field where individuals and social groups can struggle to influence and change technological design such that the very construction of technology is subject to democratic debate and contestation.

Thus, Feenberg conceives of technology as an always contested field that can be reconstructed to serve human needs and goals. Consequently, he develops a position that neither falls into a naive technological optimism, nor prey to technophobic attacks on technology. In this study, I shall interrogate Feenberg's work in developing a critical theory of technology by engaging his key works of the 1990s that spell out his general theoretical and political perspectives, and that illustrate his theory with examples of democratizing and reconstructing technologies, focusing in particular on *Alternative Modernity* and *Questioning Technology*.[3] I will indicate what I take to be Feenberg's key contributions to developing a critical theory of technology for the contemporary era and some limitations of his work.

Feenberg, the Frankfurt School and Critical Theory

Feenberg makes the interesting point that dystopic forms of media culture promoted a critical space to view technology with suspicion of the destructive aspects of nuclear technology in the aftermath of World War II and the unleashing of the atomic bomb in Japan at the end of the war.[4] Popular film, music, and discourses reflected public distrust and fear of big technology that was producing immense weapon systems, new forms of nuclear energy, and a technological society that was changing the very face of the social world while also, we were to learn later, threatening the integrity of the natural environment and even survival of the human species.

This climate helped generate a serious and positive reception of critiques of technology by theorists such as Heidegger, Weber, and the Frankfurt School, which went against the celebrations of technology in the dominant ideology of the day. Feenberg himself roots his theoretical perspectives on technology in the work of the Frankfurt School and pays homage to his teacher Herbert Marcuse in the opening chapter of his book. Marcuse took up the challenge of providing a critical theory of society in an age of affluence, arguing that technology, supposedly the source of wealth and affluence, was also a source of social domination and cultural poverty that was not meeting basic human needs for peace, freedom, individuality, and happiness.[5] Yet technology for Marcuse provides the potential to provide a better life for all *if* it is reconstructed, made accessible to the public, and aimed at the fulfillment of human needs, and not such systemic imperatives as domination, profit, and perpetuating the status quo.

Feenberg believes that Marcuse did not go far enough in linking the reconstruction of technology to social reconstruction and operated on too abstract a philosophical plane that needs concretion and further development, and he takes this challenge as his own project. Feenberg appreciates that Marcuse makes technology a political issue, subject to debate and contestation, and that it is a key constituent of the contemporary world, thus linking social theory and critique with theorizing and reconstructing technology. Feenberg, however, wants to develop what he calls "interactive strategies of change" which involve the interaction between state and corporate interests, scientists and technical designers and engineering, and the public in a complex process of negotiation and contestation over the construction and design of technologies.[6] He argues that conceiving of technology in this way opens it up as a field of negotiation, debate, and struggle over its design, effects, and ends that help democratize technology.

Rejecting dystopic positions that would simply reject and negate technology tout court, Feenberg argues that it is more productive to focus on the reconstruction of technology rather than its vilification. He claims that the post-1960s struggles have put in question absolute faith in science and technology, and the individuals and institutions that develop and implement it. With a public questioning technology, demanding changes, and in some cases carrying them out, technology is thus more flexible, transformable, and amendable to democratic debate and reconstruction than previous theories had indicated. As examples—which were fleshed out in separate studies later in his book—Feenberg suggests the ways that French consumers transformed the Minitel Videotext system from an information data base to an interactive system of communication, articulating popular desires and needs. He also discusses the ways that AIDS patients and women undergoing childbirth insisted on alteration of preexisting medical systems, as well as on the ways that the Japanese-appropriated Western technology to mesh with their own traditions and cultural and social system. In all of these cases, technology is seen as subject to contestation, reconstruction, and democratic participation, which directs it to serve human and social needs and not just hegemonic societal interests.

The subtitle of *AM* is "The Technical Turn in Philosophy and Social Theory," and Feenberg best fleshes out this dimension of his project in a chapter that engages the "technocracy thesis,"[7] and throughout the book he attempts to reconstruct the Frankfurt School and blend its perspectives with other theoretical traditions such as French postmodern theory and Japanese multicultural theory. Given Rorty's successful marketing of the notion of the "linguistic turn" in philosophy, the notion of the "technical turn" may, however, be misleading as it suggests an increasing tendency toward more technical philosophy—a trend certainly evident today, but not one that Feenberg would endorse. Instead, he means a turn to see the fundamental importance of technology, technique, and the technical in the contemporary society, and the need to develop critical and social interactionist perspectives toward this "technical turn," drawing on the most advanced theory and philosophy.

Feenberg develops his perspectives on the technical turn by criticizing the classical positions of the Frankfurt School on technology as domination (in particular Adorno and Horkheimer and Marcuse), considering Habermas's critique of the earlier Frankfurt School positions and development of his own theories, and then engaging Axel Honneth's critique of Habermas and further development of the Frankfurt School, followed by Feenberg's moving beyond Honneth toward his own perspectives. The classical perspectives of the Frankfurt School toward technology, Feenberg believes, are too pessimistic and totalizing, seeing technology largely as an instrument of domination. Habermas in turn adopts a more instrumentalist view of technology, seeing it as an instrument of technologically rational action that serves instrumental ends. Yet Habermas also sees technological rationality colonizing the lifeworld, invading areas where communication and social interaction should prevail, and thus ultimately for Habermas, in Feenberg's reading, the pessimistic Frankfurt School perspectives on technological domination continue to prevail.[8]

Habermas seeks to counter the hegemonic modes of technological rationality and instrumental reason with a notion of communicative rationality oriented toward an "ideal speech situation" in which norms of rational debate and consensus would govern concerns about technology as well as other issues of public importance.[9] Feenberg believes that this is a step in the right direction, but takes up Axel Honneth's critique that Habermas's conception of consensus and the ideal speech situation is too much of an ideal-type social myth that does not adequately take into account the issue of power and constraints on rational consensus in the contemporary world. Honneth in turn proposes that "social struggle" is the form of action taken in contests over social norms, institutions, and power, and thus develops syntheses of Habermas and Foucault that would combine analysis of power, resistance, and rational debate in adjudicating social and political issues and conflicts.

Feenberg is also sympathetic to Honneth's critique of Habermas's notion of "system," which following Parsons, Luhmann, and systems theory presents social systems as reified and depersonalized.[10] Feenberg, by contrast, wants to operate with a richer notion of the social and of organization, which sees institutions as complex conglomerates of rules and regulations, bureaucratic procedures and interests, technical imperatives, and norms and practices, all subject to contestation, debate, and reconstruction. While Feenberg is sympathetic to Bruno Latour and social constructionists who see technology and institutions as constructions imposed upon the public which dictate thought and behavior,[11] and himself introduces a notion of "implementation bias" that dictates how technology is constructed and used, he wants to make these biases and constructions subject to debate, struggle, and reconstruction, thus opening up society and technology to social transformation.

Indeed, one of the most valuable elements of Feenberg's work is precisely the way he links social transformation with technical transformation, theorizing both society and technology as fields open to social contestation and change. He also makes clear that there can be no meaningful talk of social reconstruction unless there is consideration of changing technology, of transforming its design, uses, and practices, thus linking social change with the reconstruction of technology. Feenberg is keenly aware of the central role of technology in contemporary society and that to understand and change society requires understanding and transforming technology.

Appropriating/Expropriating/Reconstructing Technology

While I am highly sympathetic to Feenberg's project and find his writings extremely useful for philosophy and social theory today, I worry that he underestimates the power of technology as a force of domination and veers too far toward an overly optimistic stance. While he rightly criticizes the classical Frankfurt School for being too pessimistic and frequently totalizing in their assault on technology and seeing it largely as a force of domination, he perhaps downplays the extent to which technology does serve as instruments of domination by societal elites. My own view is that in today's world we should see technology as both a force of emancipation and domination, holding onto the most negative critiques that we play off against the most utopian possibilities. From this perspective, it appears that Feenberg plays down too much the negative aspects of technology and is too optimistic concerning positive uses and the possibility of reconstruction.

Feenberg counters pessimistic and dystopic perspectives that technology cannot be changed, that it is the fate of the modern world to live in an "iron cage" of technological domination (Heidegger and Max Weber), with some cases studies that indicate that technology *can* be reconstructed to fulfill human needs and is subject to democratic debate and transformation. As examples, Feenberg points to a successful challenge and transformation of medical and government policy and practice as AIDS patients struggled for experimental drugs and the change of government and medical AIDS policies ensued.[12] This case also provides for Feenberg examples of how the functional imperatives of medicine treat patients as mere objects, suppressing the "caring" treatment of medicine with emphasis on "curing." AIDS patients, however, forced the medical system to address their concerns and to modify their practices accordingly.

Feenberg also devotes two chapters to the French Minitel/Videotext experiment to show how individuals have creatively appropriated existing technological systems for their own purposes and in fact restructured technology[13] and technical systems. The French telephone system initially provided a Minitel telephone/computer apparatus to each customer free of charge that would allow individuals to tap into databases to get weather and railway information, news bulletins, and other forms of information. It was intended to help enable the French to interact in a high tech exchange value and thus to aid the process of French modernization.

In practice, however, individuals hacked into bulletin boards, which were reconfigured to allow message posting, and eventually split-screen online communication and chat lines that enabled diverse forms of social interaction and connection. This expropriation shows how individuals could reconfigure technology to serve their own purposes, which may have been at odds with the purposes of those who designed the technology, as when the French used Minitel to engage in interpersonal discussion, to facilitate sexual adventures, or to promote political projects, rather than just to consume officially-provided information.

Both the AIDS and the Minitel examples show how technological systems that were devised by elites according to technical and functional requirements could be resisted by groups involved in the systems and reconfigured to better serve their own needs. Both appropriation of technical knowledge and tools for purposes opposed to their original design and implementation, and the expropriation and reconstruction of technologies and technical practices to serve countergoals and values, show that technology is more complex, flexible, and subject to contestation and reconstruction than in many existing theories and critiques of technology. This suggests the need for more multilayered theories of how technologies are introduced, implemented, and developed, and subject to subversion and reconstruction.

Moreover, as Feenberg argues in conclusion, restructuring technology and promoting technological creativity can serve as a figure for the reconstruction of society and one's way of life:

> Technological creativity is a form of imaginative play with alternative worlds and ways of being. A multicultural politics of technology is possible; it would pursue elegant designs that reconcile several worlds in each device and system. To the extent that this strategy is successful, it prepares a very different future from the one projected by social theory up to now. In that future, technology is not a particular value one must choose for or against, but a challenge to evolve and multiply worlds without end.[14]

In the final chapters of the book, Feenberg delineates some Japanese perspectives on "alternative modernity" based on a reading of the philosopher Nishida and reflections on the game of Go, which embodies values different than the Western ones of success and competition. Feenberg's point is that alternative social constructions of modernity are possible that construct different sorts and uses of technology, subjected to differing cultural traditions and aesthetic sensibilities. Thus, Nishida envisaged a Japanese modernity that combined Western modernity with Japanese cultural traditions, so that technology would be embedded in cultural and everyday practices and subject to Japanese values and aesthetics. Such a synthesis of art and technology concretizes the call for a merger of these domains by Marcuse in his conception of a new technology. For Feenberg, such conceptions relativize the Western concepts of technology, modernity, and rationality, and show that other alternatives conceptions are available. These perspectives point to a diversity of types of technology and social organization, thus breaking with the unitary and universalizing model of Western modernity and modernization theory.

In a coedited book from the same period of *Alternative Modernity*, Feenberg sketches out a conception of "subversive rationalization" which points to technological design and advances opposed to hegemonic forms of technology in contemporary Western societies.[15] While Feenberg's valorization of alternative expropriations and reconstructions of technology, of opening technologies and technical systems to debate and contestation, and to theorizing how technology can be used to serve human needs and enhance human life rather than the interests of dominant social powers is of immense importance, I have some concluding concerns. Although it is no doubt possible to challenge systems of technological domination, to reconstruct technologies, and to guide how technology will be constructed and implemented, it is also the case that technological organization of the workplace and the capitalist corporation, the state and its bureaucracies, the medical establishment, as well as the university and other institutions, are structured to a large extent by systems of technological rationality that are extremely difficult to transform and reconstruct, and even to contest.

Feenberg is certainly right that we should overcome simplistic and one-sided views of technology as either inherently an instrument of progress or of domination, but he underplays the ways that technology is currently used as an instrument of domination and how difficult it is to resist, restructure, and use it for social reconstruction. From this perspective, the more pessimistic arguments of the Frankfurt School make clear the immensity of the challenge of social transformation and the power of the dominant societal forces that create technology in their own interests. Although Feenberg emphasizes how capitalist imperatives and biases enter into the design, construction, and implementation of technology, he downplays the extent to which capital, the state, and dominant institutions themselves construct technologies in their own interests and resist alternative technologies and reconstruction.

Yet Feenberg's more activist and optimistic perspectives are more productive and useful than gloomier prognoses that only see technology as an instrument of domination. It is both useful and correct to see the social constructedness of technology and modernity and the importance of devising alternatives. Social transformation clearly requires reconstruction of technology, and it is Feenberg's merit to demonstrate both that technology is a product of social design and construction and that transforming society to make it more democratic and responsive to human needs requires reconstructing technology.

Feenberg's *Questioning Technology*

Andrew Feenberg's *Questioning Technology* is his third book in a series of studies which undertake to provide critical theoretical and democratic political perspectives to engage technology in the contemporary era. For Feenberg, technology is the most important issue of our era. It is a major constituent of contemporary society and is intimately connected with politics, economics, culture, and all forms of social and personal life. He opens *Questioning Technology* by arguing that over the past centuries democratic movements have called for debate and political control of increasing domains of social life. This process began with public debate over issues concerning the state, politics, and law under the impact of the Enlightenment and democratic revolutions. It next took the form of movements to democratize management and control of the market and the economy under the influence of Marxism and the socialist and labor movements.[16] Public debate and control over education and medicine emerged in the nineteenth century, while in this century, Feenberg suggests, democratic discussion concerning technology, its nature, effects, social management, and reconstruction is a fundamental issue for the present age.

In theorizing technology, Feenberg carries out sustained attacks on technological determinism and essentialist theories. Technological determinism follows a similar logic as economic determinism and both, Feenberg suggests, have pernicious philosophical and political implications. If the market and the economy form a quasi-natural organism, subject to its own laws and autonomy, attempts at management or control can be dismissed as interference with the natural order. Likewise, if technology is an autonomous force impervious to political control, attempting to manage or reconstruct it is either a foolish or hopeless enterprise.

Theories of technological determinism emerged after World War II which either celebrated technology's modernizing features or blamed it for the crisis of Western civilization (i.e. Heidegger, Ellul, etc.). Determinist theories thus devolved into essentialism, both of a negative and positive sort. Theorists such as Heidegger, Ellul, and their followers attributed a negative essence to technology, seeing it as a force of domination and totalitarianism. On this view, technology is a demiurge of the modern world, an autonomous juggernaut immune to democratic control or humane reconstruction, a framework, or Gestell (Heidegger), which constitutes the very structure of the modern world and lived experience. This dystopic and technophobic essentialism is contrasted to a technophilic essentialism, in which technology is characterized positively as sound control of nature, as a force of efficiency, rationality, and progress.

Technological essentialism, Feenberg notes, has given way in the contemporary era to constructivist views, which conceive of technology as socially constructed, as dependent on specific social structures and cultural values, thus robbing it of its independent force and power. Social constructivism sees the creation and development of technology as subject to contingent social factors and decisions, analyzing the specific individuals and groups who construct various technologies.[17] It rightly sees the matrix of social interests and groups that goes into the construction of technologies, but its microdescriptive and empiricist dimensions often bracket out certain overarching social imperatives and political interests. Hence, social constructivist theories separate analysis of technology from theories of society and engage in empirical description of specific technologies. Such theories abandon a more systemic and historical optic which conceptualizes technology as a key constituent of the contemporary world and which attempts to articulate and critically engage its defining features and major effects. "Thus, although constructivist sociology has placed particular technologies on the agenda in new ways, the basic questions of modernity posed by an earlier generation of theorists are rarely addressed today in terms of the general problematic of technology."[18]

Feenberg wishes to combine a form of constructivism with more systematic and socially critical views of technology such as are found in theorists like Marcuse and Foucault who analyze the links between technology and power. Such critical theorists dissect the ways that technology serves the interests of social domination, and open the space for discussion of alternative forms of technology. Feenberg links social theory and philosophy to overcome one-sided approaches which either essentialize technology or reduce it to social facts. Thus, he attempts to mediate between philosophical substantivist and social science-oriented constructivist views, criticizing philosophical essentialists, such as Heidegger, Ellul, and Habermas for their reductive, determinist, and excessively abstract views of technology. Yet Feenberg also criticizes constructionist views which solely see technology as a neutral instrument, which propose merely descriptive accounts of specific technologies in disparate historical contexts, and which renounce broad philosophical or critical perspectives.

Democratization and the Reconstruction of Technology and Society

Feenberg, then, wants to merge philosophical and sociological theories of the role of technology in modernity with reflection on actual technologies, to combine social theory and social research, philosophy and critique, analysis and reconstruction. One of his key contributions to theorizing technology is connecting philosophically oriented social theory of technology with theories of democratization. He notes that while technology is seen as a major power in contemporary society, it is often said to be incompatible with democracy. Feenberg, however, wants to demonstrate how technology can be part of a process of societal democratization and how technology itself can be restructured to meet basic human needs. In this process, technologies should be created to help produce a more democratic and egalitarian society, thus focusing on the potential for the social reconstruction of society and technology.

Rejecting all determinist and reductivist theories of technology that would ascribe to it an abstract essence, Feenberg sees technology as a contested field where individuals and social groups can struggle to influence and change technological design, uses, and meanings. In this conception, the very construction of technology is thus subject to democratic debate and contestation. Feenberg sees technology neither as determining nor as neutral, arguing that democratization requires radical technical as well as political change. He argues convincingly that there can be no genuinely democratic and progressive political change without technical change, without the reconstruction of technology, and, vice versa, no radical change of technology without democratic political change. In his view, the two are vitally interconnected and radical social reconstruction should aim at once at the transformation of society and technology.

Thus, Feenberg develops a dialectical approach to technology that perceives both negative and positive uses and effects, seeing technology as an always contested field that can be reconstructed to serve human needs and goals. Consequently, he develops a position that neither falls into naive technological optimism, nor rigid technological determinism and technophobia. Rejecting dystopic positions that would simply repudiate technology *tout court*, Feenberg argues that it is more productive to focus on its reconstruction rather than its vilification. He claims that the post-1960s struggles have put in question absolute faith in science and technology, and the individuals and institutions that develop and implement it. With a public questioning technology, demanding changes, and in some cases carrying them out, technology is thus more flexible, transformable, and amendable to democratic debate and reconstruction than previous theories had indicated.

In his major works, Feenberg succeeds in combining the articulation of theoretical and cultural perspectives on technology with concrete studies of struggles over the control and construction of technologies. In Chap. 3 of *Questioning Technology*, he suggests how the events of May 1968 in France, which he sees as the high point of the New Left, involved contestation of technocracy. This involved critique of technical control of the workplace, education, government, and culture by technocratic elites, and programs for more democratic participation and self-management. Likewise, he argues in Chap. 4 that the most progressive elements in the ecology movement—Barry Commoner is his example—call for less polluting, more sustainable technologies. Hence, the sort of environmentalism with which Feenberg aligns himself calls for the reconstruction of the technological environment and not just less production, population, and reformist practices (though these demands too have their value, as he argues).[19]

Feenberg is very skilled at marshalling examples and case studies to illustrate his theoretical and political arguments. As we have seen, in *Alternative Modernity*, he points to the retooling of the Minitel Videotext system[20] and to studies of how women struggled for alternative childbirth technologies and practices, how AIDS patients militated for alternative medicine and health care, and how Japanese critiques of technology contain conceptions of alternative models of modernity and modernization.[21] In *Questioning Technology*, he also marshals copious examples of actual reconstruction of technology to demonstrate that his project of democratizing technology is grounded in actual struggles. In these ways, he is able to counter pessimistic and dystopic perspectives that technology cannot be changed, that it is the fate of the modern world to live in an "iron cage" of technological domination (Heidegger and Max Weber). To subvert this form of determinism, Feenberg provides cases studies and examples which indicate that technology *is* subject to democratic debate and transformation and *can* be reconstructed to fulfill human needs. In his examples, technology is seen as subject to contestation, reconstruction, and democratic participation, which directs it to serve human and social needs and not just hegemonic societal interests. His examples show how technological apparatuses that were devised by elites according to economic, technical, and functional requirements could be resisted by groups involved in the technical systems and reconfigured to better serve their own needs. Appropriation of technical knowledge and tools for purposes opposed to their original design and implementation shows that technology is more complex, flexible, and subject to contestation and reconstruction than in many existing theories and critiques. Likewise, the expropriation and reconstruction of technologies and technical practices to serve countergoals and values points to the contested and constructed nature of technology and how instruments of control or domination can be transformed into tools of democratization and struggle. This sort of analysis suggests the need for more multilayered theories of how technologies are introduced, implemented, and developed, and subject to subversion and reconstruction.

Philosophical Perspectives on Contemporary Technology

I have suggested that the strength of Feenberg's approach is his integration of the development of philosophically grounded perspectives on technology with concrete studies of actual construction and reconstruction of salient technologies along with proposals for making the design and use of technology an issue of political debate and democratic politics. In the second part of *Questioning Technology*, Feenberg spells out his concept of "democratic rationalization" that includes popular participation in the adventure of technology, inserts agency into technical systems, and provides openings for the democratization of technology. In the third part of his book he turns to developing his philosophical perspectives in discussions of technology and modernity and his efforts to develop a critical theory of technology.

In polemicizing against essentialist conceptions of technology that reduce it to technique, instrumentality, *Gestell*, efficiency, and the like, Feenberg argues for an approach that "provides a *systematic* locus for the sociocultural variables that actually diversify its historical realizations."[22]

Feenberg proposes a distinction between "the *functional constitution* of technical objects and subjects, which he calls the "primary instrumentalization," and another aspect, the "secondary instrumentalization," focused on the *realization* of the constituted objects and subjects in actual technical networks and devices."[23] He argues that essentialism only offers insight into the first dimension, while the dimension of "realization" encompasses actual uses of objects, the contexts of meaning in which they are embedded, and active interaction between subjects and objects.

While Feenberg's analytic distinctions are useful in clarifying some key aspects and dimensions of contemporary technology, I think that there are some conceptual limitations in his attempt to develop an overarching philosophy of technology that will define its common characteristics over a broad range of historical contexts. There are, in fact, various levels in which a critical theory of technology can be engaged. Feenberg's earlier books developed an approach that analyzed the role of technology within a specific historical epoch, modernity, and called for a critique and reconstruction of technology in the contemporary era. *Questioning Technology*, by contrast, has a more philosophical focus with some of the analysis pitched at the high level of philosophy of technology. Analysis on this level faces the danger of excessive abstraction and philosophical projection, in which categories that are perfectly appropriate to describe technology in one historical epoch are projected onto the broader historical narrative of humanity and generalized and universalized as invariant features of the human adventure.

Feenberg's analysis of primary and secondary instrumentalization raises for me problems with philosophical theories of technology that focus on developing universalist analyses of the nature and role of technology as such in human life. Shouldn't a critical theory of technology focus more specifically and in a historicist vein on analyzing technology in a particular epoch, with special emphasis on technology in the current era, rather than providing universal perspectives on technology? While his focus was more historicist in his previous books, the concluding optic of Feenberg's *Questioning Technology* has taken a philosophical turn that strives to develop a more universalist analysis of technology that will conceptualize its invariant features in his analysis. But in so doing, he illicitly smuggles in concepts from modernity into a more general philosophical analysis.

Thus, while there may be benefits and insights generated from a more sweeping philosophy of technology that detects continuities across the vast terrain of history as well as discontinuities between historical epochs, one must be very careful in delineating the commonalities, general features, or functions that cut across historical eras. Philosophical perspectives may illuminate the trajectories of historical development, articulate both continuities and differences and discontinuities across historical epochs, criticize specific types and uses of technology from the standpoint of alternatives, and call for democratic reconstruction of technology in the present as Feenberg so persistently and eloquently has militated for. Indeed, it is the merit of Feenberg's work to disclose the specific features of many types of contemporary technology, to show how technological design can be contested and reconstructed, and to advance a democratic theory of the reconstruction of technology rather than the celebrations or dirges that characterize so much contemporary discourse on technology.

Yet I doubt whether Feenberg's concept of "instrumentalization" and his distinction between primary and secondary instrumentalizations are the best categories to adequately characterize technology throughout history in all of its diverse configurations and constellations. First, it is not clear that the term "instrumentalization" is an appropriate concept to describe the nature and function of technology throughout history. Whereas an instrumental use of technology arguably characterizes modern societies, it may be that premodern societies had more ritualistic, aesthetic, religious, or social conceptions of technology. Moreover, I am not sure that the term "secondary instrumentalization" is the correct concept for the sort of substantive analysis of meanings, aesthetic and ethical qualities, democratizing reconstructions of technology, and uses that integrate technology into specific contexts that Feenberg wants to characterize in order to distinguish certain concrete uses of technology from the more instrumental conception of technology. Feenberg has argued convincingly in his earlier works that instrumentalism is often taken as the essence of technology by many reductive "instrumentalist" theories that he strongly critiques. The term "instrumentalization" indeed seems to me to be best reserved for the dominant concept of technology against which Feenberg wants to polemicize, maintaining a link with Lukàcs, critical theory, and other critics of instrumental rationality who theorize it as a distinctive feature of capitalist modernity against which a more substantive notion of rationality is opposed.

From this perspective, technology is much more than an instrument, a term that cannot capture the rich and broad range of elements that Feenberg wants to capture in concepts like integration (of technology into everyday life), realization (of values and aesthetic qualities), and democratization (of design, uses, reconstruction). Moreover, his distinction between primary and secondary instrumentalization concedes too much to the view which Feenberg opposes by suggesting that the "primary" dimension of technology is an instrumental or functionalist one, while all other features are merely "secondary" (thus replicating the problematic distinction between primary and secondary qualities in the history of philosophy). Feenberg could argue, I would suggest, that the instrumental and multidimensional qualities of technology are equiprimordial (to use a concept of Heidegger), that design and use, meaning and function, and construction and realization are equally important in the constitution of our actual experiences and uses of technology in our social life.

Furthermore, a distinction between instrumentalization and what Feenberg calls "democratic rationalization" (to replace his primary versus secondary instrumentalization distinction) would allow all the discriminations that Feenberg wants, would continue the Frankfurt School critique of instrumental reason, and would provide a standpoint of critique to criticize not only other theories of technology, but also uses that are purely instrumentalist, abstracting from environmental contexts, values, meanings, and democratization. This distinction would capture the difference between an approach to technology that is instrumental, decontextualizing, reductive, autotonomizing, and determinist, which Feenberg wants to distinguish from an approach that is contextualizing, mediating, multidimensional, reflexive, democratic, and concrete—and which accounts for agency, values and meanings, and the actual richness of technology in everyday life that Feenberg wishes to valorize.[24]

In fact, while Feenberg's project is to develop a *critical* theory of technology, it is not clear from what standpoint of critique he is operating and how he would ground his critical perspectives. I would suggest that distinguishing between instrumental rationalization and a more democratic rationalization could provide aspects of a standpoint of critique that could be further developed in theoretical analysis and concrete studies.

In the light of the development of social media, and in particular Facebook and social networking, I worry that negative aspects and effects of new technologies are coming to overwhelm positive aspects and effects. Social media are colonizing the lives of segments of the population cocooning them into homogeneous subcultures where individuals share interests, values, and activities, that may distinguish them from other individuals. This is very positive in many ways, but in terms of news, information, and politics, it can have very destructive effects, as we have recently discovered with the shocking election of Donald J. Trump in 2016.[25] People who got their news from Facebook or other social media, which appears to be increasingly the case for many millions, found themselves consuming and circulating biased and even fake news, circulating outright falsehoods like the views that the Pope, or Denzel Washington, had endorsed Trump for President, or that the Clinton Foundation was accused of gunrunning. This has generated debates about whether Facebook has become too influential,[26] and it demonstrates how technologies can circulate propaganda, lies, disinformation, and fake news. The media have always had biases and a propagandistic and ideological dimension, but the phenomenon of fake news is tied to an overreliance and trust in technology and perhaps a narcissistic immersion in Facebook and other social networking sites. This striking example suggests that technologies can overwhelm, dominate, and transform individual behaviors, ways of assimilating and disseminating news and information, and transform their very modes of communication and social interaction in ways that can be unhealthy and even destructive for individuals and societies.

Finally, while one might argue that Feenberg underestimates the power of technology as a force of domination and veers too far toward an overly sanguine stance, I believe that his more activist and optimistic perspectives are more productive than gloomier prognoses that only see technology as an instrument of domination and destruction. It is both useful and correct to see the social constructedness of technology and modernity, and the importance of devising alternative technologies and uses. Social transformation clearly requires the reconstruction of technology, and it is Feenberg's merit to demonstrate both that technology is a product of social design and construction and that transforming society to make it more democratic and responsive to human needs requires reconstructing technology.

Notes

1. Andrew Feenberg, *Alternative Modernity* (Berkeley: University of California Press, 1995); hereafter AM.
2. Andrew Feenberg, *Critical Theory of Technology* (New York: Oxford U P, 1991[2002]); hereafter CTT.
3. Andrew Feenberg, *Questioning Technology* (New York: Routledge, 1999); hereafter *QT*.
4. Feenberg, *AM*, 41ff.
5. Ibid., 19ff.
6. Ibid., 34ff.
7. Ibid., 75ff,
8. Ibid., 76f.
9. Ibid., 78f.
10. Ibid., 81.
11. Ibid., 84.
12. Ibid., 96ff.
13. Ibid., 123–66.
14. Ibid., 232.
15. On Feenberg's notion of "subversive rationalization," see *Technology and the Politics of Knowledge*, Andrew Feenberg and Alastair Hannay, eds. (Bloomington, Ind.: Indiana University Press, 1995). This useful text provides articles on theories of technology in major theorists such as Heidegger, Marcuse and the Frankfurt School, Arendt, and others, as well as the perspectives of major contemporary theorists of technology such as Winner, Borgmann, Dreyfus, Ihde, Haraway, Longino, Latour, and others.
16. Feenberg, *QT*, viif.
17. Ibid., 10ff.
18. Ibid., 11–12.
19. Ibid., 68ff.
20. Feenberg, *AM*, 123–66.
21. Ibid.
22. Feenberg, *QT*, 201.
23. Ibid., 202.
24. See ibid., 203–208.
25. For my analysis of Trump's election and the role of social media, Facebook, and fake news, see Douglas Kellner, *American Nightmare: Donald Trump, Media Spectacle, and Authoritarian Populism* (Rotterdam, The Netherlands: Sense Publishers), 2016; *American Horror Show: Election 2016, Authoritarian Populism, and the Ascent of Donald J. Trump*.

26. David Pierson, "Has Facebook grown too influential?" *Los Angeles Times*, November 13, 2016: C1, 6.

Bibliography

Feenberg, Andrew. *Critical Theory of Technology*. New York: Oxford University Press, 1991.

———. *Alternative Modernity*. Berkeley: University of California Press, 1995.

———. *Questioning Technology*. New York: Routledge, 1999.

Kellner, Douglas. *American Nightmare: Donald Trump, Media Spectacle, and Authoritarian Populism*. Rotterdam, The Netherlands: Sense Publishers, 2016.

Kellner, Douglas. *American Horror Show: Election 2016 and the Ascent of Donald J. Trump*. Rotterdam, The Netherlands: Sense Publishers, 2017.

Pierson, David. "Has Facebook grown too influential?" *Los Angeles Times*, November 13, 2016: C1, 6.

PART IV

A Critical Response

CHAPTER 13

Replies to Critics: Epistemology, Ontology, Methodology

Andrew Feenberg

I want to begin by thanking the authors of the chapters of this book for taking the time to respond to my work. We have many points of agreement but our disagreements will be my focus here. The criticisms to which my work is subjected call for clarification and reformulation. In some cases, I have anticipated the arguments in other works to which I will refer here for the information of both critics and readers. In other cases, I will try to make my own position clearer, something I have evidently failed at in the past. The most important thing is that we are engaged in a conversation around a philosophical theme of great importance, the nature, and prospects of modernity.

Part I: Democracy

Democracy

Sassower's criticism is part of a larger argument for a postmodern social theory. His critique converges with Kirkpatrick's complaint that my

A. Feenberg (✉)
Simon Fraser University, Burnaby, Canada

D.P. Arnold and A. Michel (eds.), *Critical Theory and the Thought of Andrew Feenberg*, DOI 10.1007/978-3-319-57897-2_13

fidelity to Critical Theory narrows my vision of possible futures. Like Kirkpatrick objecting to my restricted vision of dystopia and resistance, he objects to the "binary" character of my argument in a time when the opposition of proletariat and capitalist is outmoded.

Both Kirkpatrick and Kellner are critical of my emphasis on democracy. Kirkpatrick thinks that I am too focused on democracy to pay attention to the multiplicity of values involved in the many contestations and innovations in technical life today. He and Kellner both worry that I am too optimistic about the potential of democratic intervention. Technology, they argue, is more difficult to transform than I allow, or, supposing it can be transformed, not a basis for general social transformation.

On first reading, these criticisms seemed to me to be based on misunderstandings. Although I do emphasize struggle over technology and technical systems, I have not limited myself to a binary logic, although many binary conflicts surely exist today as in the past, nor do I focus exclusively on democratic values. I have discussed struggles involving many different kinds of actors around substantive values in ecology, medicine, communication, and other social domains.[1] I do not invent the obstacles they meet. As for my putative optimism, it reflects a common interpretation of my position which I have tried to fend off without success.

I would never argue that democracy always yields good decisions, or that it is easy to change technology and technical systems. What I do argue is that democratic interventions are sometimes possible and have had significant beneficial consequences. If holding out the possibility of progressive change is "optimistic," then I am an optimist, but no more so than the activists who engage in progressive struggle. Kellner's doubts about the ability of political protest to change technology raise a difficult issue. It is possible, as Marcuse argued in *One-Dimensional Man* that there is no escape from technocratic domination. Perhaps it is useful to be reminded of this today.

Kirkpatrick thinks that my focus on technocracy is outdated. This criticism gets us closer to the underlying issue, although I do find it odd in a period of triumphant conservatism in which pseudo-science justifies deregulation and corporate overreach. However, Kirkpatrick sees in the emphasis on technocracy a lingering fidelity to outdated dystopian fears of technology, the sort of thing manifested most clearly in Heidegger.

He shrewdly analyzes a fracture in my approach between dystopianism and pluralism. The dystopian vision is based on the notion that human potentialities are suppressed by a technocratic hegemony. This sets up the restricted dualistic framework that he and Sassower reject. On the contrary, a pluralistic theory of politics affirms the multiplicity of values and futures available to a society that plays freely with technological possibilities, without, however, the promise of a revolution. Kirkpatrick concludes, "Feenberg's distinctive ontology of technology…inhibits his development of the idea of radical technical politics."

To explain my response to this critique, I must go back to the original impulse behind the critical theory of technology. As a student, I was schooled in Heidegger and Marcuse's ontological or substantive critique of technology, but 10 years later I found myself working with two innovative technical institutes, a medical foundation engaged in clinical research and a research foundation that developed the first online education program.[2] Experience with technical systems called into question my reliance on a deterministic ontological critique, and yet I could see clearly that the issues raised by the critique had certain validity: human beings were indeed suppressed as they were squeezed into the conventional technical forms from which we were trying to free them.

Marxism was a tempting alternative. It placed responsibility for the problems with the capitalist employment of technology in pursuit of profit. But the usual Marxist formulations overlooked the role of technical design in the alienating consequences of technology. The ontological critique could not be so easily dismissed.

My experience with technology led to a new critical strategy which I recognize now as a kind of metaphoric "demythologization" of ontological critique. The conundrum could be resolved where the influence of capitalism was located not just at the level of particular business decisions, but at the level of a general, culturally secured design strategy exhibited throughout industrial society and having some of the main effects deplored in the ontological critique. That would leave a space for the sort of innovative multiplicity to which Kirkpatrick appeals without unrealistic postmodern assumptions about the transcendence of technocracy and capitalism in a new form of open society.

The role of democratization in my theory is not based primarily on anti-dystopianism or normative considerations, but on an analysis of the practical conditions of gaining compliance under different social arrangements. The point is not that the pursuit of profit is irrelevant (nor even that it is immoral) but that it presupposes management, a highly centralized control of labor in disregard of conflicting interests of workers and communities. It would be nice for the owners of enterprise if workers spontaneously devoted themselves to the welfare of the firm, but that has not been the historical experience of capitalism; hence the centralization of control through management and deskilling in most if not all business enterprises.

Management uses its control to design the production system in ways that reproduce that control. Thus the system can only be changed by a democratization of ownership and control to interest subordinates in the welfare of the firm. I call the centralized control characteristic of capitalism and imitated by communism "operational autonomy," and argue that in the twentieth century it was extended far beyond the factory as techniques of management were exported to every area of social life.[3]

Although I argue that gaining compliance takes two main forms in modern societies, the power exercised from above or cooperation organized from below, I need to make two qualifications to avoid misunderstanding. First, other forms of action coordination take place in some institutions, for example, collegial administration, and there are also spontaneous orders such as markets. But these alternatives are not consistently relevant to the large-scale organizations of modern societies. Second, I do not claim that power is absolute. Obviously, there are limits. Capitalism negotiates those limits with governments and either explicitly with workers or implicitly with the labor market. But within those limits, it has operational autonomy and uses its relative freedom of decision to reproduce its autonomy through technical arrangements.

Centralized control has substantive consequences for subordinates. Some of these consequences are intrinsic to the very nature of hierarchy, for example, barriers to trust and truthful communication. Others are due to differences in priorities. For example, the priority given health and safety on the job tends to be lower where management organizes production in pursuit of profit than where those who do the job face the consequences of their own decisions. This is, of course, an empirical claim that should be supported by research on such matters as the different negotiating priorities of unions and management. But university professors can verify the difference from their own experience, if not in the domain of health and safety, then in aspects of job satisfaction stifled by ever more heavy-handed management.

The dystopian logic persists in this new form without ontological presuppositions and as a contingent effect of social organization. I disagree with Heidegger that technology is inherently responsible for the dehumanization and abuse we see around us in modern societies. I also disagree with Kirkpatrick that technology bears no inherent dangers. That seems wrong for practical if not for ontological reasons.

The hope I place in democratization is also free of ontological presuppositions. I believe that technology and social institutions can serve a wider range of interests where a wider range of opinions must be taken into account in the design process. Note the qualification: "can serve," not will serve. I rely on the notion that rational self-interest plays a significant if not always predominant role in modern social life; I do not think that an unreasonable assumption.

Kirkpatrick seems unwilling to accept even this vestige of dystopianism. He claims that the "technical illusion" has "evaporated." The deterministic understanding of technology that legitimates the dominant hegemony is no longer credible. Are we truly so advanced? I doubt it. In the field of technical development I know best the quasi-dystopian logic of my argument still holds. Online education is fraught with tension between those who would like to employ it in the interest of traditional educational values and corporations and corporatized universities that employ it to deskill education and to replace faculty with machines. Nothing has changed since Andrew Ure complained in the 1830s about uncooperative workmen (read: professors) and looked to the "union of capital and science" to eliminate them.[4]

Nevertheless, I do agree with Kirkpatrick that there is no predetermined endpoint of progress. The concept of "potential" cannot deliver an a priori theory of progressive goals. We must reconstruct the idea of progress in a future-oriented manner in the light of the ongoing tradition of human struggles for freedom and fulfillment.[5]

This brings me to the pluralistic aspect of Kirkpatrick's analysis. He notes my interest in Laclau and Mouffe's theory of political articulation. They intended to free political theory from essentialist concepts such as class interest. Instead, politics was said to be totally contingent, dependent on identification with symbols. Articulation describes the nature of struggles and alliances, but Laclau and Mouffe seem to deny altogether the relevance of the objective facticity of those in struggle and of the technical environment. Surely these facticities play some sort of role. Kirkpatrick may or may not agree with this critique; he draws on Laclau and Mouffe principally to argue that the development of technology is contingent and unconstrained by any predetermined interest or ideology. He says, "The diffusion of digital technologies has encouraged the development of diverse cultures of experimentation, dabbling, reconfiguring, sabotage, and so on." I agree but do not see this as incompatible with my demythologized version of dystopianism.

What Kirkpatrick calls my dystopianism stems not from Heidegger but from my reading of the concept of rational domination in Lukács and Marcuse. In Lukács's version, the logic of dystopia is called "reification," but he argues that its grip on the social world is contested in dereifying social action. On Kirkpatrick's terms, this would mean that oppressive social forms are not fate but opportunities for innovation. Kirkpatrick's argument for cultural experimentation is implicit in this notion of a new type of democratic agency, disturbing the rational systems that order modern society.

Environmentalism

During one summer of my college years, I worked for Barry Commoner. Commoner wrote an environmental best-seller *The Closing Circle* that is still an interesting read. He ran for President on the Citizens Party ticket in 1980 and while he did not get many votes, many voters got his message. It was his educational work and that of other scientists that gave the American public concepts and explanations for environmental issues they confronted in their experience. Their growing understanding of the environment eventually led to important legislative and technical changes.

Although I never took a class with him, I count Commoner as one my most influential teachers. I stayed in touch and wrote about his work.[6] I learned from him that technology is not neutral, that its design is contingent on economic interests, that the relation of design to human needs is also contingent, and that environmentalism should be based on the rational self-interest of the sort Marx expected to motivate class struggle. Arnold offers a good summary of my current views on the environment which still hold to these early convictions, although as he points out, I evaluate the role of culture differently from Commoner.

Arnold argues that my "views of democracy remain at a very abstract level." He would like a clearer outline of the system of governance of an ecologically oriented society, and an account of the international framework within which environmental goals might be achieved. These are valid concerns but they have not been my preoccupations. I have instead reacted to the tendency among environmentalists to focus on the state, that is, on legislation and regulation. We are so used to the idea that politics is about legitimate procedures and government that it is not easy to understand a conception of politics that decenters these institutions.

I do not believe in a "constitutional" or "procedural" solution to the problems of the environment that relies primarily on a reform of political governance. This has to do with the fact that environmental problems are essentially technological. As such, they respond to standard procedures implemented by expert personnel. So long as those procedures and those personnel are committed to environmentally unsound practices, the government can only paste band-aids over the wounds of nature. That is surely better than nothing but a deeper change is needed, a technological change, and that requires implementation by technical personnel. Ultimately, they must translate public concerns into technical codes for significant change to take place. What is the best way to achieve this?

Like Arnold and Dryzek, I doubt the virtues of what Dryzek calls "administrative rationalism," that is, technocratic expertise in a representative democracy. Both interests and ideology militate against deep changes in the existing system, and administrative rationalism offers no way to address these obstacles.

We inherit a technological system built under a form of capitalism that was much better insulated from public pressures than it is at present. The technologies deployed in earlier times were less dangerous compared with the toxic chemicals, radiation, and massive release of CO_2 associated with the industry today. The collision of publics with technologies we now witness is an aspect of a general transformation of the industrial heritage in response to the enlargement of the range of interests it must serve. This transformation is costly for industry and conflicts with commonplace business strategies. The resistance is tremendous and continues even as corporations claim the "green" mantle.[7]

Further obstacles result from the role of technical expertise. The experts work with representations of nature based on natural science. The boundaries of their disciplines do not always correspond to everyday experience with nature, with the result that dangerous side-effects may be overlooked. Furthermore, the scientific basis of much modern technology supports the technical illusion, the belief that what is right because it is (scientifically) rational.

Arnold reminds us of Sunstein's discussion of the distortions introduced into popular views by bad "heuristics." These are sometimes invoked to disqualify the public from involvement in technical matters. But experts have bad heuristics too. Some of the most consequential are a bias in favor of the views of those with wealth and power, conforming to political pressure, sticking to technical traditions against new evidence, and discounting disturbing reports by lay people of harmful side-effects. Without the check of public pressure, experts often fail us.

These considerations explain why we cannot count on administrative rationalism to rush to identify and change harmful technical practices. It was not automotive engineers or Ford executives who initiated the struggle against smog, but citizen complaints. Environmental movements are most often set in motion from below, from the human victims of environmental abuse. The lived nature they encounter signals problems ignored by those in charge.

I argue that it is a mistake to propose an ideal model of this environmental politics. It is more useful to look at actual struggles. On that basis, I identify three modes of democratic intervention into technology: controversies such as protests and boycotts, creative appropriations or reinventions such as user innovation and hacking, and innovative dialogues through such procedures as citizen juries. These modes of intervention are directed at changing the understandings and practices of technical experts, either through influencing them directly or indirectly by influencing the government or corporate officials who set policy.[8] As environmental crises worsen, these interventions seem likely to carry more and more weight.

I agree with Arnold that the solution is mutually correcting dialogue between the lay public and experts.[9] But the dialogue need not be "deliberative"; it may also be conflictual and involve lawsuits, boycotts, and protests. Either way, the message eventually gets through. This happens now ever more frequently as technical issues intrude more forcefully on daily life.

Today, lay involvement is not recognized as a normal aspect of technical development. Were it so recognized, technical politics would be an acknowledged informal mechanism in the structure of technical governance, much like freedom of speech and assembly in the traditional public sphere. This is another way of conceiving democracy suited to modern societies that depend on reified systems they must also challenge. But there are major obstacles to this development, principal among them the exorbitant power of capital in resisting change, and regulation where profits and autonomy are threatened.

Arnold also raises the question of international environmental protection which I do not address. However, a response is implicit in my work. Technical disciplines and codes are international in scope and can carry environmentally sound approaches to nations around the world.[10] This is the achievement of the 1989 Montreal Protocol, which protects the ozone layer by requiring many small technical changes on a global scale. Promoting change at this level seems to me the way in which international agreements and institutions can play a role. It would be enormously helpful if the World Trade Organization enforced basic environmental standards as vigorously as it protects business, but I see this as likely only under extraordinary public pressure.

The Internet

The question of democracy in advanced capitalism centers more and more on the impact of the Internet. Greaves situates my work in relation to two different Marxist approaches to that question. One of these approaches affirms the neutrality of technology and overestimates the effects of the new forms of communication made possible by the Internet. On this view, the Internet is to be understood primarily as the material basis of a new revolutionary class. The other approach makes the opposite mistake by treating the Internet as an integrative mechanism and overlooking its political potential altogether. A dialectical philosophy of technology is necessary to make sense of the Internet. Greaves argues that I provide such a philosophy. He explains, "Technical networks exert forms of control over users. The inverse is however also true. Technologies allow user manipulation in a variety of ways. This relationship is the core of Feenberg's dialectic."[11]

Greaves' chapter raises the question of how my work relates to its Marxist background. I have borrowed much from Marx and the Western Marxist tradition but there is a fundamental divergence. The divergence has to do with the role of the industrial proletariat in social change. The central political role Marx assigned this class is explained by the specific stage of capitalist development in his time rather than by the nature of capitalism as such. The factory is no longer the center of the capitalist universe. Technology and technical modes of thought have spread everywhere, bringing in their wake social relations of domination and democratic resistances.

The facts today differ considerably from those in Marx's day but they remain susceptible to a dialectical treatment that identifies the continuing contradiction between those in control of technical systems and those subordinate in those systems. This new dialectic differs, however, in important respects from the contradiction of capital and labor. It is not revolutionary, but nor does the traditional concept of reform adequately encompass the stakes in the struggles. Huge changes of civilization are implicated in environmentalism, new roles for women, and struggles over the Internet. The extent to which these changes will ultimately prove compatible with capitalism is unknown.

Greaves shows the difficulty of conceptualizing these new struggles over the Internet in Marxist terms. These struggles for control of this new medium of communication have only just begun. They pit users engaged in communicative applications, including political communication, against corporations attempting to channel attention toward commercially profitable targets. Governments too appear as adversaries in some situations, spying on dissenters and repressing dissent. These struggles play out in attempts to influence regulators and in dramatic public actions such as the famous internet "blackout" of 2012, protesting repressive legislation.

Greaves argues that this case can be understood in a modified Marxist framework of the sort I have applied to technology. Technology is still at the center of the theory, as it was for Marx, but the theory must change as the place of technology in the totality changes.

Part II: New Worlds for Old?

Technoscience

Sassower "wonders what is at stake in adhering to a line of demarcation between science and technology."[12] Isn't it time to throw out that distinction and admit that science and technology are united in a single phenomenon called "technoscience?" I stick to the old distinction and argue that two things are at stake.

First, and most simply, the term "technoscience" does not encompass most technology. It is true as Sassower points out that in many contexts it is difficult to distinguish scientists from technologists, but it is not at all difficult to distinguish scientists from truck drivers, secretaries, construction workers, and fast food cooks. All these workers encounter technology in the lived world rather than through science. We need to be able to talk about their labor without always dragging science along for the ride. The considerations that govern the lives of these workers in their technical involvements and encounters with nature are simply too different from the activities of scientists and the nature science describes to identify them all with the term "technoscience."

Second, there is a strategic problem that first became clear in the Lysenko Affair in the Soviet Union when political interference in research went beyond such extrinsic matters as policy proposals and funding decisions to select the "true" theory from among competing scientific alternatives.[13] Anyone who reads the current news in the United States should know that this is not an outdated concern. However, while direct interference of this sort is disastrous in the case of research, it goes on every day in the life of corporations making decisions about technology.

Consider climate change, the current subject in dispute. The science involves research that uses a great deal of technology but that has as its product research papers, a representation of nature. At the same time, corporations are trying to respond to regulatory changes by adopting new technology that will have effects on nature with which we engage in everyday life, for example, how hot it will get on summer days. To the extent that the government mandates such technological changes, it obliges corporations to apply the science, a perfectly legitimate governmental function. But if the government now decides to throw out the science in favor of an alternative preferred by a politician or an oil company that would be a scandal no philosopher of "technoscience" would want to defend.

This said, I do agree with Sassower that the concept of technoscience is useful, but I would limit its use to particular domains in which the overlap of scientific and state or commercial activity is preponderant. I have written about this issue using cold fusion as an example.[14] Contrary to custom in the scientific community, cold fusion was first presented at a press conference that announced simultaneously a scientific discovery and a commercial development project. The cold fusion experimental apparatus was presented as a commercial prototype, a technology. The experiment failed to convince other scientists and the project was dropped, but the coincidence of experimental and commercial prototype is more and more common today, especially in the biological sciences. This justifies the use of the concept of "technoscience."

Transhumanism

It is difficult for me to respond to Michel's chapter because we believe in two completely different sets of facts. He is apparently convinced that what the National Science Foundation (NSF) calls the "convergence" of nano- and biotechnology with computing will lead to a transhuman future. Total technology will so transform human beings that the question of democratic control will no longer be relevant. In the fully realized *Gestell* democracy will be technologically obsolete.[15]

I concede that the NSF has great authority, but in my view, it displays the bias of corporate capitalism in choosing these particular technologies as its model of "convergence." I would have proposed ecology in which many sciences converge not to replace us but to save us. That would suggest a different future in which democracy might play an essential role.

In any case, I doubt that the vision of the NSF will come to pass. Of course, robots will continue to evict humans from certain jobs and perhaps drivers from the drivers' seat. Computers may someday make those pesky voice menus on the telephone more efficient. But I am convinced that the larger posthumanist projection is hype intended to extract funding from government and corporations. Scientists are furious about fake science in cases such as tobacco "research," but they show a surprising tolerance for fakery when it concerns the promise of their legitimate colleagues' work. We should view their most amazing claims skeptically.

Consider the huge sums spent on artificial intelligence and the rather slender results to date. True, powerful computers can beat chess and go masters, but why is this considered a step toward a posthuman future? Human services and professions such as journalism, education, and law can certainly use smarter computers to improve performance, but their logic is qualitatively different from the calculation of moves in a game. The meaning of chess and go as human activities is irrelevant to the case, which concerns only the supposed "purpose" of play in a society obsessed above all with winning.[16] Are horse races less interesting because cars can drive faster than horses can gallop?

The case of socialbots is revealing. In a restricted environment of online text, bots can fool human beings into thinking they are real people. They may pass some sort of "Turing test," but this is not an advance but rather a regress. It shows the extent to which people have adopted a machinic identity, or at least have such conventional expectations of each other online that human beings are easily imitated by a computer program.[17] The movie *Her* made this clear by depicting the man seduced by his "intelligent agent" as a passive lump compared to lively Scarlett Johansson, his virtual lover. Let's hope this is not our future!

Her repeats the cliché that computers lack only feeling to be like humans. In sum, where humans are active, computers will replace them, while pathos remains our last redoubt. This grants too much to algorithms; it ignores the equally significant deficit of the computer in many human activities not dedicated to efficiency. The goal of technological advance is increased efficiency, not such goals as challenge and creativity. Perhaps as technology succeeds in one domain after another in achieving "posthuman" levels of efficiency, humans will attach ever more value to other achievements of which they alone are capable. Far from being replaced, we may hope that humans become more "human."

Part III: Lukácsian Marxism

Reification

Kavoulakos offers a good explanation of the theory of reification and defends my interpretation of it against Honneth's.[18] Reification is a "form of objectivity," that is, an a priori of experience characterizing the historical epoch of capitalism. It is based on common practices such as market exchange, which shape both a social world and a corresponding subjectivity. As such it is more than a matter of ideology or personal belief. As Lotz argues, it crosses the line between epistemology and ontology. But in saying this, Lotz thinks he is criticizing my supposedly subjectivistic account of reification. This he attributes to an over-reliance on early texts of Marx at the expense of his later theory of the commodity. Yet I say in *The Philosophy of Praxis* that "Lukács relates Marxism to the problem of rationality through a revolutionary philosophical interpretation of *Capital*." The following pages of that book develop this point in some detail.[19]

Kavoulakos describes my position accurately: "The capitalist form of objectivity [reification] can be understood as a cultural pattern that forms the relation between subject and object, based on calculative-instrumental rationality." I have concretized Lukács's concept in the notion of "social rationality," that is, a form of rationality modeled on scientific principles and, in its application to society, shaping a reified social world. I list three such forms corresponding to administrations, markets, and technologies. These forms are classification under rules, mathematical equivalence, and precise measurement and optimization. The institutions implement these forms and are legitimated by their supposedly rational basis.[20]

Significantly, Lukács does not propose an alternative organic form of objectivity as a socialist response to reification. The alternative is a dialectic of reification and dereification. The reified institutional structures and technological designs depend on practices and can be changed. This involves, as Lotz argues, a transformation in the relation of particular moments, particular institutions and designs, to the social totality. Such transformations begin at the grass roots, in the practices that ground the society. They cannot eliminate reification once and for all. The imposition of cultural patterns based on calculative-instrumental rationality will persist in any modern society. But the institutions can become more fluid, more responsive to protest, and innovation from below.

This is the contribution of socialism to democracy as it extends citizen agency to rational systems. Lukács writes, "the world which confronts man in theory and in practice exhibits a kind of objectivity which—if properly thought out and understood—need never stick fast in an immediacy similar to that of forms found earlier on. This objectivity must accordingly be comprehensible as a constant factor mediating between past and future and it must be possible to demonstrate that it is everywhere the product of man and of the development of society."[21]

Some of Lawson's criticisms might be answered by reference to these clarifications of the concept of reification. He seems not to recognize that the notion of reified social rationality is a sociological observation, not an argument. He thinks I should justify it as though it were my own personal view. But I do not attempt to justify it because I do not think it is justifiable; it is simply the existing state of affairs in a reified capitalist society.

Lawson argues that capitalist institutions attempt to achieve the status of closed systems resembling those of scientific experimentation; and he concludes that this is an "error" that invalidates social sciences such as economics. I agree with Lawson up to a point, but the implications are not only epistemological. The failures of economic science reflect the nature of the real economy with which human beings, not just economists, must live.

Lawson wonders if I endorse the attempts by administrations and economists to achieve closed systems and scientific isolability. On the contrary, I argue that the attempt to impose a reified form has human consequences that come back to haunt the institutions as resistances. Society cannot be made to conform to the natural scientific idea of nature. The concepts of closed systems and isolation are useful precisely for showing just how impossible the technocratic dream really is. That is the essence of my understanding of democratic technical politics, not an objection to my argument.

In sum, resistance is not a side-effect of cognitive errors but a social structural consequence of the reified nature of capitalist society. Lawson has my position backwards because he sees reification as a cognitive issue, when in fact, I argue that the cognitive dimension is a highly mediated reflex of social forms such as commodity exchange and operational autonomy. This does not directly contradict Lawson's views but goes beyond them, as Lotz might say, from epistemology to ontology.

Instrumentalization Theory

The instrumentalization theory is where all these themes are supposed to come together. I have tried various formulations but seem unable to make clear where I am going with this theory.[22] Both Lawson and Kellner misunderstand it no doubt due to my lack of clarity. I will try to correct that now.

Lawson has an interesting proposal that relates my philosophy of technology to critical realism in the philosophy of science. Critical realism claims that science works with closed systems isolated from the surrounding world. This is what makes it possible to discover laws of nature. Lawson argues that the primary instrumentalization involves a similar isolation of phenomena. I think what he means is that technical artifacts must be relatively isolated from disturbances in the environment to function, much like an experimental apparatus.

This misconstrues the nature of the distinction between primary and secondary instrumentalization. I intend the distinction to be analytic. It shows technology to involve two different forms of thinking, "if…then" causal thinking and hermeneutic thinking. These forms converge on a common outcome, a design. There is nothing mysterious about this combination in everyday practice: we turn on the porch light (causally) in order to help (hermeneutically) our guests find our house ("It will be the one with the light on"). Craft labor is similar: I take a photograph of a meaningful object while manipulating optical technology according to various causal rules of thumb.

The case gets more complicated with modern production which is at the intersection of the nature of natural science and lived experience and meanings. The quasi-scientific thinking of engineers is combined with symbolic considerations of meaning. These constitute cross sections in the concrete unity of the final product. For example, to satisfy the marketing department with a sporty design, the car must have a low hood. That requires the engineers to select a certain type of engine, and so on. True, the marketers and the engineers think differently and belong to different divisions of the company, but their efforts are seamlessly combined in the car they design together.

This is what I intended with the instrumentalization theory, but Lawson treats my analytic distinction as a real distinction. He seems to believe that technology, like an experimental apparatus, is not symbolically shaped. But even the experimental apparatus is shaped by theoretical assumptions from which it cannot be "isolated" as it is from causal perturbations.

Unlike an experimental apparatus, technology belongs to the wider social world. It not only needs protection from causal perturbations, but it must acquire a meaning and a role in social life. As such it is always affected by secondary instrumentalizations which intervene in the design process in order to accommodate technical insight to society. Lawson does not see that secondary instrumentalizations are there from the very beginning but thinks of them as mere "afterthoughts." It is true that secondary instrumentalizations may intervene after an artifact is designed and released, as in the case of environmental protest and reform. Perhaps this can be seen as an afterthought, but the technologies environmentalists engage with always already combine primary and secondary instrumentalizations.

Kellner's generous description of my work concludes with a critique of the instrumentalization theory. He argues that the theory "illicitly smuggles in concepts from modernity" in the attempt to provide a universal theory of technology. The very term "instrumentalization" reeks of the modern differentiation of value spheres. He would prefer a theory limited to modernity in which the contrast between technocracy and democracy replaces concern with the essence of technology.

I propose a "structural" response to the question of the essence of technology. I believe in the existence of human universals such as language, music, and gait, which depend on a genetic endowment all humans share. The ability to think technically is one of these universals. But the concrete realization of these universals is necessarily culturally particular. They have no reality apart from some specific cultural enactment. I argue that there is an underlying logic to each universal that is combined with other logics and varied in the course of history. In the case of technology, the basic logic includes an abstractive ability that allows affordances to appear and the foresight to anticipate the consequences of actions involving those affordances.

As Kellner points out, the understanding of the technical itself varies widely in different cultures. Not all of them would distinguish technical action from other forms of action as we do. To that extent, the instrumentalization theory is indeed a "modern" theory, but then so are all the theories we can propose. Even cultural relativism presupposes a concept of culture unique to modernity and so is itself ethnocentric. The instrumentalization theory does not claim to reveal the way other cultures understand themselves but rather describes what they do as we can best understand it. Given its basis in common humanity, mutual understanding through translation ought to be possible even in an encounter between radically different cultures. The peasant who plants after checking the moon for signs of the weather does not meet my scientific criteria of technical explanation, but his idea of the technical is similar enough to mine in form that we can easily communicate, convince each other, or agree to disagree.

I am reminded of a story told by a composer who brought a piano to a village in Bali in the 1930s. He performed Western music for one of the local gamelan players who had never heard our music before. "Nyoman, a Balinese musician for whom McPhee played unspecified specimens of Western music on an upright piano, complains that it sounds 'like someone crying.... Up and down, up and down, for no reason at all.'"[23] The gods would not be pleased to hear human beings weep. Clearly, the meaning of "music" was radically different for the Westerner and the Balinese, but that did not prevent them from recognizing each other's music as music. And there is no reason a Westerner should hesitate to analyze gamelan music in terms of Western concepts such as harmony.

The instrumentalization theory attempts to describe the general structural features that are particularized culturally and to outline the relations between technical thinking and these cultural contextualizations. I suspect this is a fundamental disagreement, but I will concede to Kellner that the terms "primary" and "secondary instrumentalizations" which I have used to describe these two dimensions of technique are confusing. In my recent book *Technosystem: The Social Life of Reason* I substitute the terms "causal" and "cultural functionalizations." I am not sure this is an improvement from Kellner's viewpoint, but I am hopeful that it will clear up some of the confusion that results from a chronological ordering of what I intended to be an analytic distinction.

Two other chapters have interesting things to say that relate to the instrumentalization theory. Lotz argues that exchange value is the form of appearance of use value under capitalism. He writes, "The concept of use-value should not be conceived as something 'external' or outside the system. Use value, in other words, takes on a specific form under which it exists in capitalist social formations." This Hegelian formulation means that use value is always already mediated by exchange value, that is, by the requirements of commodification. The general point could be extended to precapitalist societies where use values are mediated by cultural traditions rather than the market. In any case, needs do not emerge and are not satisfied in a purely naturalistic framework; society always intervenes between the natural basis of human life and the means of satisfaction.

This structure resembles the instrumentalization theory. Needs and the use values that satisfy them stand in the place of the primary instrumentalization and the exchange relation that shapes them resembles the secondary instrumentalization that accommodates them to a particular culture. As in the instrumentalization theory, the two levels are distinguished analytically, not ontologically. Recognizing this, I have generalized the instrumentalization theory to cover not only technology, but also commodities and, recently, administrations as well. I call the complex of technologies, markets, and administrations the "technosystem."[24]

Belu applies the instrumentalization theory to the study of in vitro fertilization. Her example brings out the relevance of certain aspects of Heidegger's thought to the critical theory of technology. Here indeed dystopianism figures as Kirkpatrick charges, but this appropriation of Heideggerian themes is more complicated than he credits.

As I read her description, in vitro fertilization performs the primary instrumentalization through decontextualizing and reducing the egg while autonomizing both the medical personnel and the patient and positioning the medical institution commercially. So far, this resembles the Heideggerian *Gestell*. But this primary instrumentalization is structured by a secondary instrumentalization that involves various recontextualizations: criteria of selection, the constitution of a "normal" pregnant woman, professional codes of behavior and communicative practices, and physician and consumer initiative. Belu concludes that insufficient attention is paid to the physical and mental stress and harm of the procedure, especially when it fails as it often does. The prevailing secondary instrumentalizations do not address some of the problems raised by the intervention. The critique suggests ways to improve the process; it is not purely negative as one might suspect from the Heideggerian background.

This indicates an important difference between critical theory of technology and Heideggerian dystopianism. Critical theory is constructive and seeks to identify possibilities of change. It is not content with *Gelassenheit*, the "letting-be" that Heidegger recommends in the face of technology. Where he advocates a "free relation" to technology, critical theory calls for the transformation of technology into the basis of a better world.[25]

On Nature

Vogel's chapter challenges my understanding of nature. I cannot reply in detail to his long and complicated argument. Instead, I would like to explain my own position which I think differs from the one Vogel criticizes in significant ways. In the process, I hope to clarify ambiguities that do not seem to me to go to the core of my position. I responded to earlier formulations of his argument in *The Philosophy of Praxis* but evidently, I have not convinced him.[26] I will try again here to situate the key disagreement which results in so many other apparent differences. It turns out we are not so far apart as Vogel thinks.

In *The Philosophy of Praxis*, I defend what Gramsci called "absolute historicism" according to which social practice is the ontological foundation of other regions of being, including nature. The point of this speculative concept is to establish the universal reach of reason and of rational practices, or, more accurately, to deny the existence of regions of being inaccessible to rationality in principle. To this end, Lukács, like Gramsci, affirms the ultimate sociality of nature. They believe this is necessary to combat the growing irrationalism of the bourgeoisie and, in the case of Lukács, reified social and economic thought. In this, they continue the work of Enlightenment under the influence of German idealism. This is a difficult position to explain, and Lukács's *History and Class Consciousness* offers a problematic defense of it. My own position is based on a development of that position in the course of which I attempt to resolve the difficulties.

The ambiguous social role of natural science and nature is the basis of the problems. In the first chapter of his book, Lukács picks an argument with natural scientific methods. These methods exemplify the reified rationality of capitalist society; they involve quantitative measurement of facts and ahistorical laws. They are, he says, appropriate for the study of nature but not for the study of society. "When the ideal of scientific knowledge is applied to nature it simply furthers the progress of science. But when it is applied to society it turns out to be an ideological weapon of the bourgeoisie."[27] This distinction seems to imply that the reification of nature is valid, beyond criticism, while society is only reified in appearance, its truth revealed to critique.

That is the conclusion drawn by critics who note that Lukács excludes nature from the dialectic. He writes that "the crucial determinants of dialectics—the interaction of subject and object, the unity of theory and practice, the historical changes in the reality underlying the categories as the root cause of the changes in thought, etc.—are absent from our knowledge of nature."[28] Another passage states that "no single act can eliminate reification in one blow. It means that there will be a whole host of objects that at least in appearance remain more or less unaffected by that process. This is true in the first instance of nature."[29]

To compound the confusion, he later states that "Nature is a social category."[30] Although he quickly qualifies this initial lapidary claim to apply only to "what passes for nature," it is often read as contradicting his earlier exclusion of science from the dialectic and dereifying critique. In that case, a "successor science" not based on reification might supersede our current knowledge of nature after the revolution. This leads to the bizarre conclusion that the practical subject of social life also constitutes nature. Could it be that Lukács wants us to believe stars and dinosaurs are not really "out there" as independent existents but are posits of society? With all this to contend with it is no wonder that Lukács's position is differently interpreted and often dismissed.

The problem is not just Lukács's but is proper to Marxism. Natural science appears independent of ordinary social forces and the nature it describes has no special place for human beings and their achievements. On the other hand, there are obvious similarities between scientific ideas such as "the survival of the fittest" or mechanical models in physics and corresponding social phenomena such as competition and machinery. One might conclude that despite its appearance of autonomy, science, like political and religious beliefs, depends on social forces, even perhaps economic interests. Nevertheless, Marx did not assimilate science to the ideological superstructures, and neither does Lukács. To do so would be absurd in a modern society in which science has become a force of production. Accordingly, Lukács recognizes the distinction between science and ideology and affirms the cognitive value of science and its methods in the study of nature.

But he also rejects positivist and empiricist explanations for the success of science, that is, epistemologies in which science is based on a direct relation between the individual mind of the researcher and natural phenomena, unmediated by society. Lukács argues that science belongs to society in a unique manner different from ideology. What is that manner? The intermediary position he stakes out is complicated and requires far more argument than he provides in *History and Class Consciousness*. As a result, he seems either to be saying that nature and society are separate realms of being or that nature "in-itself" is unknowable.

A long line of critics starting with Deborin and Rudas in the 1920s question whether Lukács's position on science and nature is compatible with Marxism. Kavoulakos in this volume raises the question of whether Lukács satisfies the requirements of absolute historicism. Vogel replies to both questions in the negative and raises the same questions with respect to my work.

It is interesting that problems with *History and Class Consciousness* identified nearly a century ago preoccupy contemporary critics such as Vogel today.[31] These early criticisms have returned in the context of the rise of ecology, a science that crosses the lines between society and nature in new ways Lukács did not anticipate. Vogel complains that despite his ambitious philosophical rhetoric Lukács confined his critique of reification to society and showed only that social institutions are founded on collective practices, while leaving nature to one side unaccounted for, indeed, permanently reified. In that case, Lukács's historicism would be limited by an apparently alien being, nature, given over to the tender mercies of industrialism.[32] The ontological distinction between nature and society is incompatible with environmentalism.

Vogel goes on to claim that like Lukács I downplay the ontological significance of labor. This is the source of the dualistic ontology. He proposes instead a unified ontology based on the interaction of human beings with nature in labor. Labor is the true "praxis" which I ignore in my work.[33] He rejects the distinction I make following Lukács between our understanding of nature as an object of labor and the nature of natural science. He insists that the nature of natural science is "built" just like the artifacts that populate the social world. Scientists work, too, in the construction of their facts and theories. I agree, but the object of that work is concepts and data, not natural objects. Stars are not transformed by astronomers as trees are by lumber jacks. Vogel seems to deny the difference between theory and practice. This has implications, which I discuss below.

The critique overlooks one significant passage in *History and Class Consciousness* that explains the entire process of reification through the imposition of the commodity form on labor. That passage discusses the construction of abstract labor through the wage relation and the transformation of production technology. Lukács argues that as labor is subsumed under capitalism it loses its concrete, qualitative form in craft and takes on a reified form suited to industrial production. Space and time are subject to strict measurement, and the workers' labor is reduced to quantitative units of abstract labor power.[34] The reified form of labor turns out to be the basis of the concept of nature under capitalism.

This passage already suggests an answer to Vogel, but his emphasis on labor has made me aware of an ambiguity in Lukács and in my own appropriation of his theory. It is true that the terms "society" and "history" appear frequently in his book in juxtaposition with "science" and "nature." The full implications of the theory of reification are not drawn explicitly and tend to be obscured by an apparent continuation of the neo-Kantian argument for a distinction between the social and the natural sciences and their objects. It is clear that some such distinction is necessary, but the way in which it is made matters.

I will argue here for a position similar to Lukács's answer to his critics in an unpublished text called *A Defence of History and Class*

Consciousness: Tailism and the Dialectic.[35] In that text, he attempts to correct his earlier omission by acknowledging the "metabolism" of humanity and nature in the labor process. In the course of this discussion, he distinguishes implicitly between two ideas of nature, the theoretical nature of natural science, and the practical nature of the labor process. The starting point for understanding the former is the latter, "…the real historical interaction of subjective and objective moments of development…in its living interaction."[36] Lukács thus registers the fact, obvious from a Marxist standpoint, that society cannot be described without its technology and the nature implicated in technical operations. In the course of this discussion, he also argues that all reality, including nature, is known through historically evolving categories. In capitalist society, those categories are reified as is their source in the labor process.

Lukács's reference to "living interaction" marks the distinction between theory and practice, and on that distinction depends the further distinction between nature as a scientific concept and nature as an environment within which human beings act. Unfortunately, Lukács does not develop the latter concept, but we must do so to make sense of his theory.

Here is an analogy that may help to clarify the distinction. Peasants create a socialized nature through their labor. They do not represent nature scientifically but have their own ideas about how it works and how it can be transformed, ideas based on experience and tradition. If they use scientific ideas it is in the context of this vernacular notion of lived nature. In terms of that idea, human beings and nature are truly joined in labor, mutually necessary for the production of the land and its harvest.[37]

A painter who paints this nature creates a representation of a different type from the peasant's idea of nature. The painter's representation does not transform its object materially but leaves it as it was before. The painter's tools work on the representation, not its object. Where the peasant engages with nature both at the level of meaning and through a physical transformation, the painter engages with the landscape only at the level of meaning. The painter's physical labor transforms the canvas, not nature.

But at the level of meaning, the painter does have an effect on the world. The painter imposes an aesthetic form of objectivity on the natural environment that makes of it a "landscape." Augustin Berque explains that landscape, in this sense, is a concept and an aesthetic experience that emerges in history as a consequence of urban development.[38] The work of painting involves a kind of "isolation," as Lawson might argue, that has specific historical conditions. Those conditions include the labor of the peasants who unconsciously shape the landscape into an aesthetic object. Painting is thus historically contingent even though it depicts nature as it is in itself. Substituting natural science for painting, the analogy brings my position closer to Vogel's than the one he attributes to me in his chapter. The dualism, if such it be, divides not society and nature but two forms of nature, the lived nature encountered in labor and the theory of nature elaborated by natural science.

This is my understanding of how Lukács intends to incorporate labor in *A Defence of History and Class Consciousness*, but this text does not address his concept of absolute historicism explicitly. For that concept to make sense the two natures (or two perspectives on nature) must be related to what Marx called "the demands of reason." The critique of Lukács shows the importance of working out these relations even if this carries the argument beyond the point where he left it. This will require a better understanding of the concept of reification as it relates to the concept of reason.

Absolute historicism is a Marxist version of rationalism. Lukács explains the Enlightenment's universal claims for reason as both a defense against the mystical claims of religious and feudal ideology and an ideological reflex of reification which, under capitalism, knows no bounds. The universalism of bourgeois philosophy is a reflection of the universal extension of the commodity form. To that form correspond theoretical principles such as quantification and technical manipulability which are the basis of the search of modern science for facts, laws, and technical control. If Marxism is to defend the universal claim of reason, it must do so on a different basis.

Kant explained the correspondence of the principles of scientific reason and the structure of the objective universe in terms of constitutive subjectivity. The world of experience was shown to depend on forms such as space, time, and causality, imposed by the mind. These forms constitute the world in a transcendental act of consciousness. This is the "identity of subject and object" that validates the universal powers of reason and eliminates mystery from the world and, with it, the precapitalist forms of authority grounded on faith.

But this rationalism is troubled by two limitations. On the one hand, the individualism of capitalist society is reflected in the identification of constitutive subjectivity with the individual mind. This creates an implausible disproportion between the constituting and the constituted. The individual subject can only constitute its own perceptions, not their source in the thing-in-itself reflected in perception. On the other hand, the supposedly universal form of scientific rationality meets an explanatory limit in the social world. Facts and laws do not give an accurate description of society because society is not a collection of independent things such as science finds in the natural world. Rather, society is based on practices, dynamic processes of human relations that take on the reified form of things as a socially necessary appearance under capitalism. It is as though bourgeois social science mistakes freeze frame images for the continuous flow of the movie.

Lukács argues that Marxism overcomes these limitations by shifting the paradigm of rationality from natural science to social critique. The transcendental constitution of nature by the subject in Kant's scheme is replaced by the Marxist theory of society in which constitution is a real process of production of the social world through action in history. The constituting subjectivity is now social rather than individual. As such it can really produce social objects through its practice, not simply imagine a mythic individual constitution.

But what is the place of natural science in the new scheme of things? Does the critique of reification apply to scientific reason in the same way it applies to social beliefs and institutions? The unfamiliarity of Lukács's contemporary neo-Kantian philosophical context is what gives rise to the difficulties we have encountered in answering this question. If that context is ignored, it seems that he confronts the dilemma sketched with the dueling quotations at the beginning of this section: either science is a nonsocial, immediate relation to the facts, or it is ideology and the revolution will transform nature. But in context a third possibility appears.

As we have seen, Lukács uses the neo-Kantian term "form of objectivity" to describe reification. Although Lukács describes reification as a sociological or cultural pattern, as a form of objectivity it is also a precondition for understanding a realm of objects, hence, a form of rationality. Such realms include aesthetic objects, physical objects, persons, and so on. Each realm has a certain "meaning" attached to it through which its contents are recognizable and understandable. In Lukács's usage, forms of objectivity belong to historical epochs and determine their perspective on existence. Reification is thus a way in which the world reveals itself to subjects under capitalism. It is one way among others for the real world to present itself in its truth. It is not an ideology, but an opening. Even in its relation to social reality which, as we have seen it misconstrues, reification reveals a fundamental truth: the fact that society is a human creation.

The concept of form of objectivity enables Lukács to historicize natural scientific knowledge without reducing science to ideology. The socially determined categorial structure of science determines what we know as nature. And that knowledge is not an illusion to be contrasted with nature-in-itself. On the contrary, Lukács follows neo-Kantianism in evacuating the thing-in-itself as a hypothesis which merely represents the contingency of knowledge. Its hypostasis as a world "behind" the known world and forever beyond reach corresponds to a false ideal of knowledge as an immediate and nonsocial predicate of a subject. The in-itself simply refers to the possibility of error or progress in knowledge, not to a transcendent world we can never know. The nature we experience, like the idea of nature elaborated by natural science, can thus be called "real" without qualification.

But something strange is going on here. In Kant and most neo-Kantian philosophy forms of objectivity are attributed to the subject. But Lukács locates the form of objectivity in social reality. It is not primarily conceptual but practical. The real abstractions that transform labor under capitalism institute reification. This practical reification enacted in the most fundamental relation of human beings to nature makes modern science possible by opening a theoretical perspective on natural phenomena.

Having historicized knowledge, Lukács must raise the question of science under socialism, but he is remarkably cautious. Might socialism introduce a new non-reified form of objectivity that would alter the basis of science and its concept of nature? Lukács mentions the possibility of categorial change but appears to limit its impact on science to the emergence of historical perspectives on nature as in the theory of evolution. Other aspects of reification such as quantification and the construction of facts and laws go unmentioned and uncriticized. He thus appears to accept as permanent aspects of the reification of natural scientific nature. He says that in any case, the evolution of science is a matter for scientific research rather than for critique or speculation.[39] Perhaps the reason for his caution has to do with the dialectic of reification and dereification discussed above. Lukács does not seem to know just how far socialism can go in dereifying the social world, much less natural scientific nature.

Given the practical reconstruction of the categories and these reservations concerning the future of science, does Lukács's version of absolute historicism achieve the "identity of subject and object?" In less Hegelian language, can he show that dialectical social critique and practice validate the universal claims of reason against any and all mysteries? It is clear that this issue is not the one Vogel raises concerning the materiality of labor. In practice, that materiality may just as well confirm as cancel reification whatever ontological conclusions one draws. Instead, the question of the reach of rationality must be posed at the level of the form of objectivity, of meaning. At that level, the question concerns whether the practice can transform social meanings or whether those meanings remain caught in the "immediacy" of reification.

Let me return to my analogy with painting to explain this. The status of meaning in this sense is more complicated in the case of society than in landscape painting and natural science. Unlike the actual landscape, which is indifferent to the painter and vice versa, or the reified form of the objects of natural science, the reified form of social institutions is lived by human beings who suffer under it. The cows in the picture have no real relation to their portrayal and so cannot be harmed by the aesthetic form of objectivity imposed on them, at least not directly. Similarly, actual atoms and molecules are not affected by their scientific representation, or if they are in certain interpretations of quantum mechanics, they can't complain. But human beings cast in the reified form of objectivity of capitalism are subjugated and abused. This is first exemplified in the effects on the health of workers and the land Marx discusses in *Capital.* The environmental movement responds to the damage suffered by nature today in its effects on human life.

Social movements arise from the narrow restrictions on the individuals' moral and physical existence.[40] Their lived experience comes into conflict with the reified forms. The movements aim at particular reforms, but the precondition of agency is a break with reification which they invalidate practically. In so doing, they change the relation of particular moments of social life to the totality, or, in less Hegelian language once again, they change the function and meaning of the institutional realities that they challenge. Lukács argues that this process of challenging reification is at the basis of the Marxist dialectical understanding of society. It is a dialectical reason in a practical form, overcoming at the level of practice the false naturalization of social reality imposed by capitalism.

A theoretical perspective that fails to transcend reification in thought cannot explain this underlying dynamic of its objects and must encounter such movements as irrational outbursts. This is why the method of the natural sciences is inappropriate for social sciences and policy making. We have seen this all too often as experts reassure the public that its water is safe, its cars are pollution-free, and a little radiation can do no harm. The condemnation of the environmental movement as ideological still rings in the ears, even these many years after industry has, for the most part, changed its tune and decided to co-opt it.

In his considerations on resistant practices, Lukács makes a surprising leap from social theory to absolute historicism. He argues that the irrationality of social crisis and struggle against reification occupy the philosophical locus of the Kantian thing-in-itself with respect to bourgeois social science. Crisis and struggle emerge from the uncomprehended reality behind (capitalist) knowledge which is limited by reification to surface appearances.[41] When the crisis strikes, it is as though the thing-in-itself manifested itself in experience, a Kantian impossibility but an all too frequent challenge to the universal claims of reason under capitalism. Thus, in the case of social knowledge, appearance, and reality stand in a real relation, not simply the ideal relation characteristic of natural science.

The astonishing "fall" to earth of Kant's transcendental system has huge implications for rationalism. Where the Enlightenment sought to validate reason against irrational superstition, Lukács renews the struggle for reason in opposition to its own reified form. That form posits a world subject to an uncontrollable fate—the law of the market—still more pitiless than the will of the gods. Bourgeois rationalism thus ends up in resignation rather than proving the powers of a liberated humanity as its original program proposed. Reason can only achieve true universality through a self-critique that validates the struggle against reification. "The demands of reason" require this remarkable advance in reflexivity.[42]

This brings rationalism into the twentieth century and forms the background to the early Frankfurt School's Critical Theory. The critique of instrumental reason is an application of Lukács's argument as resistance declines in the face of new and more effective forms of domination. Reason persists as a critique of the limits of reification even in the absence of practical struggles verifying its claims. Instead, the critical theorists point to the pathologies of domination that emerge as reification tends toward totalization. Among those pathologies is the hatred of racial and foreign enemies. The most characteristic contributions of Critical Theory emerge at the intersection of scapegoating and reification. Those who claim that it is outdated need only read the newspapers to verify its continuing relevance.

Let me conclude now by situating my argument with respect to Vogel's. I agree with him that labor is the original scene on which the relation of subject and object plays out. But he ignores the complexity of reification. Under capitalism, routine labor has an ambiguous status. It is, for the most part, commodified, alienated, reified, and yet lived by the individuals who make and use its products as their world. They are united with nature in the privative mode Marx described in 1844 with the concept of alienation. Lukács shows that this reified form of alienation has spread beyond labor to the entire social world. It now affects the administration, law, the media, science, and art. All these domains are subject to various dereifying struggles.

The spread of reification to other social institutions besides the production system and the decline of the proletarian movement open a space for new forms of radical struggle. In this context, the distinction between science and lived nature takes on a different aspect. The limitations of the reified natural scientific concept of nature appear in technological applications oriented exclusively toward profit and power. Narrow specialization and biased measurements of efficiency lead to designs with noxious side-effects experienced by ordinary people. Their resistance to these side-effects invalidates the claims to the rationality of the technocracy and renews the logic of dereification.

Environmental struggle is the prime example of this dynamic. It dereifies technologies and the associated concept of lived nature. In doing so, it orients science toward different problems and technological applications without challenging the reification of nature in scientific research. This is an example of the new form of democratic agency in which dereification consists. It does not abolish reification in every domain but places it in a fluid relation to social action. The instrumentalization theory analyzes the implications at the level of technical design and social communication. The experience of nature enters technology and technical systems through secondary instrumentalizations responding to the "living interaction" of humans with the environment. Insofar as the project of Lukács's absolute historicism still makes sense, it is through such struggles.

Notes

1. I will take this opportunity to reference writings that supplement my comments here. With respect to this comment, see Andrew Feenberg, *Questioning Technology* (New York: Routledge, 1999), Chap. 3; Andrew Feenberg, *Alternative Modernity: The Technical Turn in Philosophy and Social Theory* (Los Angeles: University of California Press, 1995), Chap. 5; Andrew Feenberg, *Transforming Technology* (New York: Oxford, 2002), Part II; Andrew Feenberg, *Between Reason and Experience: Essays in Technology and Modernity* (Cambridge, MA: MIT Press, 2010), Chap. 5. For another critique and response to my discussion of democracy, see Gerald Doppelt, "Democracy and Technology," and Andrew Feenberg, "Replies to Critics," in Tyler J. Veak, ed., *Democratizing Technology: Andrew Feenberg's Critical Theory of Technology* (Albany: SUNY Press, 2006), 85–100, 196–199.
2. Years later I wrote several articles presenting ideas formed by these experiences. See Andrew Feenberg, et al. "The On-Line Patient Meeting." *Journal of Neurological Sciences* (1996) 139, 129–131; *Alternative Modernity*, Chap. 5; Andrew Feenberg "Building a Global Network: The WBSI Experience," in L. Harasim, ed., *Global Networks: Computerizing the International Community*, (Cambridge, MA: MIT Press, 1993), 185–197; *Transforming Technology*, Chap. 5; See also Andrew Feenberg, "Encountering Technology," http://www.sfu.ca/~andrewf/encountering.pdf, accessed Jan. 31, 2017.
3. Andrew Feenberg, *Critical Theory of Technology* (New York: Oxford, 1991), 28–29, 79–80.
4. Andrew Feenberg, "The Online Education Controversy and the Future of the University," *Foundations of Science*, (2015). doi:10.1007/s10699-015-9444-9, 1–9. See also Edward Hamilton, *Technology and the Politics of University Reform: The Social Shaping of Online Education* (New York: Palgrave, 2016).
5. Andrew Feenberg, *Technosystem: The Social Life of Reason* (Cambridge, MA: Harvard University Press, 2017), Chap. 8.
6. Andrew Feenberg, "Barry Commoner," *Encyclopedia of Environmental Ethics and Philosophy*, B. Callicott and R. Frodeman, eds. (New York: McMillan Reference, 2008), 159–161; *Questioning Technology*, Chap. 3. On environmentalism, see also *Between Reason and Experience*, Chap. 2. For another discussion of my approach to environmentalism and my response, see Andrew Light, "Democratic Technology, Population, and Technological Change," and Andrew Feenberg, "Replies to Critics," in *Democratizing Technology*, 136–152, 199–204.
7. Feenberg, *Between Reason and Experience*, Chap. 2.

8. Feenberg, *Questioning Technology*, 120–129.
9. Feenberg, *Technosystem*, Chap. 7.
10. Feenberg, *Between Reason and Experience*, Chap. 6.
11. Feenberg, *Technosystem*, Chap. 4.
12. Feenberg, *Transforming Technology*, 170–175.
13. Loren Graham, *What We Have Learned about Science and Technology from the Russian Experience* (Stanford: Stanford University Press, 1998).
14. Andrew Feenberg, "Technoscience at the Fork," in Jay Foster, ed., *Continental Philosophy of Science*, (New York: Bloomsbury, 2017).
15. For another discussion of the issue of transhumanism and my response, see Simon Cooper, "The Posthuman Challenge to Andrew Feenberg," and Andrew Feenberg, "Replies to Critics," in *Democratizing Technology*, 19–36, 188–191.
16. Feenberg, *Alternative Modernity*, Chap. 9; Andrew Feenberg, *American Go E-Journal*, Monday June 13, 2016, http://www.usgo.org/news/2016/06/rational-play-the-master-of-go-vs-alphago/, accessed Jan. 30, 2017.
17. Andrew Feenberg, "Preface: The Human Being in the Age of Mechanical Reproduction," in *Socialbots and Their Friends: Digital Media and the Automation of Sociality*. R. Gehl and M. Bakardjieva, eds. (New York: Routledge, 2017.)
18. See also Andrew Feenberg, "Rethinking Reification: On Alex Honneth's Reading of Lukács," in T. Brewes and T. Hall, eds. *The Fundamental Dissonance of Existence: New Essays on the Social, Political and Aesthetic Theory of Georg Lukács* (New York: Continuum Press, 2011), 101–120.
19. Andrew Feenberg, *The Philosophy of Praxis: Lukács, Marx and the Frankfort School* (London and New York: Verso, 2016), 2–3, Chap. 4, especially 84–86.
20. *Between Reason and Experience*, Chap. 8.
21. Georg Lukács, *History and Class Consciousness*, trans. Rodney Livingstone (Cambridge MA: MIT Press, 1971), 117–118. For my discussion of this passage, see Andrew Feenberg, *The Philosophy of Praxis: Lukács, Marx and the Frankfort School* (London: Verso, 2016), 117–119.
22. A recent attempt is Sara M. Grimes and Andrew Feenberg, "Critical Theory of Technology," in *The SAGE Handbook of Digital Technology Research* (London: Sage, 2013), 121–29. I offer another recent account in *Technosystem*, Chap. 6.
23. Robert Craft, *Bali-Hai*, http://www.nybooks.com/articles/1991/10/24/bali-hai/, accessed Jan. 30, 2017.
24. I have generalized the instrumentalization framework to commodification in *Between Reason and Experience*, 170–171. For further discussion see Paul B. Thompson, "Commodification and Secondary Rationalization,"

and Andrew Feenberg, "Replies to Critics," in *Democratizing Technology*, 112–153, 201–202.

25. For other discussions of my position on Heidegger and my response, see Trish Glazebrook, "An Ecofeminist Response," Iain Thomson, "What's Wrong With Being a Technological Essentialist: A Response to Feenberg," and Andrew Feenberg, "Replies to Critics," in *Democratizing Technology*, 37–52, 53–70, 192–196.
26. Feenberg, *Philosophy of Praxis*, 133–143.
27. Georg Lukács, *History and Class Consciousness*, trans. Rodney Livingstone (Cambridge MA: MIT Press, 1971), 10.
28. Ibid., 24 n6.
29. Ibid., 206.
30. Ibid., 10, 24n, 234.
31. Abram Deborin and Nikolai Bucharin, *Kontroversen über dialektischen und mechanistischen Materialismus* (Suhrkamp: Frankfurt, 1968). See also Paul Burkett, "Lukács on Science: A New Act in the Tragedy," *Historical Materialism*, Volume 21, Issue 3, pages 3 – 15, and my reply "Fracchia and Burkett on *Tailism and the Dialectic*: A Response," *Historical Materialism*, Volume 23, Issue 2, 2015, pp. 228–238.
32. Feenberg, *The Philosophy of Praxis*, Chap. 6 reviews the argument.
33. As Vogel points out "praxis" is an ordinary German word. I spent hours playing from a book of Piano Praxis as a child. I adopt the word following Gramsci's usage for a body of theory but see no reason to employ it where we have a perfectly adequate English translation. "Practice" does appear in my text and can be found in the index. I define philosophy of praxis at several places in the early pages of my book, for example, *The Philosophy of Praxis*, 3–6.
34. Lukács, *History and Class Consciousness*, 87–92.
35. Georg Lukács, *A Defence of History and Class Consciousness: Tailism and the Dialectic*, E. Leslie, trans. (London: Verso, 2000).
36. Ibid., 101.
37. I have tried to clarify this relation to nature with a phenomenological conception of "world" uniting subject and object. Belu shows in her chapter that this concept, developed by Heidegger in *Being and Time*, describes the human subject as co-constituted with its tools, hence as a laboring being. In that context the antinomy of nature and society dissolves in the sense that the laboring act corresponds to potentials in nature realized in production and use. This is an essential relation in a historically evolving context. As Marcuse discovered in the early 1930s when he read Marx's *Manuscripts of 1844*, this phenomenological concept of world is strangely similar to the early Marx's concept of nature, but requires a basic reformulation to take into account the alienating effects of

capitalism on human beings and their labor. See Herbert Marcuse, "New Sources on the Foundations of Historical Materialism," *Herbert Marcuse: Heideggerian Marxism*, R. Wolin and J. Abromeit, eds. (Lincoln and London: University Nebraska Press, 2005); Andrew Feenberg, *Heidegger and Marcuse: The Catastrophe and Redemption of History* (London: Routledge, 2005), 122–126; *The Philosophy of Praxis*, Chap. 3.

38. Augustin Berque, *Thinking Through Landscape*, trans. Anne Marie Feenberg-Dibon (New York: Bloomsbury, 2014).
39. Ibid., 117–118. For a further discussion of this issue, see *The Philosophy of Praxis*, 137–143.
40. Lukács, *History and Class Consciousness*, 166.
41. Ibid., 179.
42. Ibid., 126–131.

CHAPTER 14

Appendix: Interview with Bruna Della Torre de Carvalho Lima and Eduardo Altheman Camargo Santos

Bruna Della Torre de Carvalho Lima and Eduardo Altheman Camargo Santos

Q: Many scholars have been writing about the possibilities of a new praxis today. Examples one remembers are Fredric Jameson, Slavoj Zizek, Stanley Aronowitz, etc. Some say that communism is no longer an alternative and others state that the major challenge today, against the neoliberal "there is no alternative" empire, would be to reopen your political imagination to the idea that there is, in fact, an alternative and to substitute the hegemony of the management politics to the great politics again, as Gramsci would put it. And you just wrote a book about the philosophy of praxis and the

This interview was first published in Portuguese in the vol. 7, n. 2 (supl.) of Revista Idéias (http://www.ifch.unicamp.br/ojs/index.php/ideias).

B. Della Torre de Carvalho Lima (✉) · E. Altheman Camargo Santos
University of São Paulo, São Paulo, Brazil

D.P. Arnold and A. Michel (eds.), *Critical Theory and the Thought of Andrew Feenberg*, DOI 10.1007/978-3-319-57897-2_14

relation between theory and praxis in Marx's, Lukács's, and the Frankfurt School's critiques. What is your stand about the relation between theory and praxis today? Is it still possible to be a Marxist in an orthodox sense?

AF: There are two questions here: what is Marxist theory, and what is the relation of revolutionary theory to political practice. Let me reply in order.

Lukács claims that Marxist orthodoxy is a question of method. In that sense, yes, it is possible to be an orthodox Marxist. But it seems that no one agrees with Lukács on this point. Everyone thinks Marxist orthodoxy requires believing specific factual claims made by Marx or his authorized followers, such as Lenin. Marxism would be a theoretical tradition or "research program" based on their thinking. I have to say I am ambivalent about this understanding of Marxist theory. So much has changed in the social and intellectual world since the consolidation of this tradition in the years following the Russian Revolution that only a fairly drastic revision makes sense today and whether it can still be called Marxist is certainly in question.

Let me give an example on which I am currently working. How do we place Foucault's contribution in relation to Marxism? His emphasis on the role of rationality in modern society, what he calls "power/knowledge," contrasts with the critique of the irrationality of capitalism in the Marxist tradition. And yet there is a connection since Marx inspired Weber's theory of rationalization and Lukács's related theory of reification. Can we work out all these connections in a way that remains within the Marxist tradition? Perhaps, but at the price of significantly stretching its boundaries. Not everyone will go along with such an operation.

As for the second question, there is a difficulty: what is the revolutionary practice to which theory is supposed to relate? Is it the labor movement? The environmental movement? A "new communist party," or an old Trotskyist one? I would argue that all the various social movements for progressive change are the practice to which our theory must relate, but I would also deny that these are revolutionary movements in anything like Marx's sense.

We are faced with a peculiar situation. The *Communist Manifesto* outlines reforms conditioned on the success of the revolution, many of which have been implemented under capitalism without a revolution. This is a continuing process. The revolution no longer appears as the condition *sine qua non* for major progressive reform. It was once possible to argue that these premature social reforms, such as social security, co-opted the class and integrated it to capitalism. This was the great argument of the Frankfurt School. But by now, so many similar reforms have transformed capitalism and daily life, and revolutionary expectations are so diminished, that that argument is no longer convincing.

After the fall of the Soviet Union, do we know what we mean by socialism? Do we really want to say that the civil rights movement, the women's movement, the environmental movement, movements for changes in medical care, and so on integrate capitalism at the expense of a revolutionary potential assumed to slumber in the masses? Isn't it more likely that these movements have set capitalism and the masses on an evolutionary path the future of which is unforeseeable?

Theory can still play a role in this situation. We need to reconstruct the idea of socialism and the dynamic of the reform movements, their forms of protest and communication, and their interactions with technical and administrative structures; all this is a hint at political innovations that might take on increased importance in a new form of socialist society.

Q: You say that the differentiation of traditional and critical theory was directly possible because of the revival of the dialectics that Lukács, Korsch, and Bloch took on. How do you see the relationship of the critical theory with other sources, especially the non-Marxist ones, such as phenomenology and Heidegger, or Weber's critique of rationalization, or Freud's psychoanalysis for instance?

AF: Critical Theory was transformed by Habermas and cut off from its roots in the earlier tradition of dialectical Marxism. This would be the main new source to which dialectical Marxism relates today. It is not productive to simply ignore Habermas. His theoretical innovations are important, but they seem tied up with a specific historical moment that has passed just as thoroughly as the moment in which the Frankfurt School first developed its approach. I have tried to learn from Habermas while restoring the central concern with technology of the first generation. Technology disappears from Habermas's work fairly early in what seems an unacceptable concession to technocratic authority. Unfortunately for Habermas, he elaborated his position before the environmental movement thoroughly discredited the technocracy and proved that ordinary people can intervene productively in the orientation of technical development. My work is predicated on the implications of that situation for the critique of technology.

Habermas's distinction of "system" and "lifeworld" points to the duality of modern social organizations and artifacts. On the one hand, they exhibit a rational form while, on the other hand, they reflect social choices and values. But Habermas tends to reify this distinction as though some institutions exemplified one side of the duality and other institutions the other side. In reality, it cuts across all institutions and artifacts. Working out the "double aspects" of modern social life can draw on theoretical resources found in Heidegger and Lukács. This is the general program of my work.

Q: Lukács saw the crisis as a possibility of somehow cracking the reified appearance of a non-contradictory society, and therefore, it presents itself as an opportunity of reuniting theory and praxis. Do you think it is still possible to sustain this thesis? Especially, if we think about the 2008 global economic crisis?

AF: I think all the social institutions and technologies of modern capitalist society are reified and all the protest movements they provoke are engaged in dereifying practices. Lukács explained reification and dereification in terms of the relation of social form and content. The categories of political economy are a form imposed on a social content that overflows them. That content is essentially the life process of the proletariat which cannot be contained in economic functions. Today, many forms are imposed by bureaucracies and other social institutions on human lives with similar consequences.

To understand how this relates to your question, we need to clarify what Lukács meant by "reuniting theory and praxis." I do not think Lukács meant that praxis is reunited with theory when it implements Marx's idea of a socialist revolution. Nor does this concept have anything to do with theory subordinating itself to the party or the class. That is taking everything Lukács intends backward. The point is not that practice conforms to a preconceived theoretical idea or vice versa, but something much more subtle and complicated. And especially interesting!

Theory for Lukács is a prolongation of class consciousness, and class consciousness itself must be understood as a kind of practical method, a way of understanding and approaching practical problems. Marxist theory is based on the methodological approach that derives from the practice of the working class. As Marx (more or less) says in *The 18th Brumaire*, theory employs the same means to solve in thought the problems practice solves in life. The unity of theory and practice depends on the class dereifying itself and re-signifying itself and its place in society. That, in turn, enables theory to achieve a dereified understanding of society as a whole.

The nearest approximation to this in recent political history is the function of consciousness in the women's movement. A new self-understanding is tied to new practices and gives rise to a new theory. There is, of course, a difference: Class consciousness has the power not only to re-signify the worker's identity but also to overthrow the society that depends on workers obeying all the obligations that flow from that identity. None of the recent social movements have this system-changing potential.

Q: In your book, you say that one of Lukács's problems understanding the generalization of the reification process is that he had neither a psychoanalytical inspired theory of reification nor a media theory. What rule do you think the media have concerning the Critical Theory of the Frankfurt School? What do you think is the place of the culture industry, to quote Adorno and Horkheimer's concept, in the present blockage of praxis?

AF: The first commercial radio broadcasts in the USA begin in the 1920 s, and television only reaches the population as a whole after World War II. It is normal that Marx underestimates the impact of powerful technologies that only emerged generations after his death. The Frankfurt School was contemporary with these technologies and appreciated their importance. What they witnessed was the construction of a new culture based not on Marx's cultural environment in which opinions were shaped in a struggle between religious institutions, communication among a literate public, and artistic production, but rather by broadcasting under the control of corporations and politicians. The culture industry is the take-over of the public sphere by business and propaganda. There can be no doubt that this is the primary factor in the debasement of culture and politics today.

I realize there are desperate academic holdouts for some sort of virtuous reciprocity between audiences and broadcasters (it's called "cultural studies"), but who can take this seriously in the age of climate denial and Trump? Please, get a grip! The media really are the dominating force in promoting conservative political movements, just as those old-fashioned Marxists of the Frankfurt School claimed long ago.

Q: In a talk you gave at the International Critical Theory Conference of Rome in May 2016, you spoke about the importance of Lukács for the Frankfurt School. The title of your presentation was provocatively "Why students of the Frankfurt School will have to read Lukács." It seems that nowadays, there has been a revival of the Frankfurt School that does not include either Marx or Lukács in their syllabuses. How do you see that and what are the traps such a reading of Critical Theory could lead to?

AF: I suppose that by a revival of the Frankfurt School, you mean some combination of Habermas, Honneth, and narrowly scholastic readings of Benjamin and Adorno. I have to say I treasure all these deviations from a radical political understanding of the Frankfurt School. But I do so as a scholar, that is, as someone curious about scholarly issues. The larger question of the significance of this revival of the Frankfurt School is a different matter. Yes, we can learn all sorts of important and interesting things from the current trends in the interpretation of the Frankfurt School, but we need more than these trends can offer. We need to see the relevance of this tradition to the political struggles going on today. The best way to achieve this is to go back to Marx and Lukács, the most important influences on the politics of the Frankfurt School thinkers.

Q: There also seems to be nowadays a general critique of the Frankfurt School that tries to denounce the alleged elitism and Eurocentrism present in their theories. Such denunciations would render their critique not radical enough, obsolete, and a consequence of their German *Bildung*. And it seems to lead to a complete negation of their work. How do you see the radicalism of Critical Theory today?

AF: I am tired of this trash talk on the academic left. Too many people are pushing their "products" by negative advertising. Whatever the flaws of the Frankfurt School, it has a lot to contribute. I attended one of Foucault's last talks in the USA. Before reading his talk, he opened with a 15-minute peroration praising the Frankfurt School and denying that he had abolished the subject, a then commonplace reading of his work. In fact, he said, the Frankfurt School had anticipated some of his most important ideas and his whole effort consisted in understanding the subject. I took this to be Foucault thrashing his American epigones who thought he could be used to beat up on their intellectual adversaries, those terrible paleo-Marxists of the Frankfurt School.

Q: Let's discuss specifically Adorno's critique of the New Left. Do you believe his appraisal would have been the same if he had lived say in Paris, where workers and students truly marched together, calling for a general strike, highlighting "imagination in power," the need to "demand the impossible" and so on? Don't you believe his critique of the desperate actionism of the New Left has a lot more to do with the specific situation in Germany and in the USA, where isolated groups could not unite and form a larger group to contest capitalism?

AF: Adorno's critique of the New Left is embarrassing. This is seminar room critique, at the level of the great political challenge he experienced when the Frankfurt Institute seminar room was occupied by his students. Of course, he was upset, but you are supposed to be able to deal with this sort of thing if you are a revolutionary thinker. When several female students mocked him by exposing their breasts in his seminar he should have asked them to come closer, not run away outraged by their lack of academic piety. The poor guy...

I find it hard to understand Adorno's attitude. Marcuse engaged with students, many of whom were just as narcissistic and confused as Adorno's students, but he criticized from within the movement, not as an outsider diagnosing a disease. Even though the German student movement may have been particularly nasty, surely Adorno was aware of the heroic struggles over civil rights in the USA and the vast popular movement in France in 1968. Why was he so provincial in his evaluation of the New Left, as though his own miserable experiences were of world-historical importance? It is a mystery to me.

Q: On the other hand, Marcuse became known for being the guru of the New Left, the so-called *maître à penser* of the barricades. But don't you think this is also a reductionism of Marcuse, in the sense that his theory is never really discussed? His political activism seems to count much more and to be his only contribution to Critical Theory. How do you see these different rhythms of philosophy and politics, as Habermas put it?

AF: The underestimation of Marcuse today is a sign of the degeneration of the academic left. Some serious thinkers such as Bernard Stiegler continue to debate respectfully with his thought, but a lot of others simply rip off themes from his work without attribution and count on the ignorance of their audience to grant them credit for ideas he first explored.

There are two amazing facts about this situation. Before *One-Dimensional Man* sold 300,000 copies, almost no one in the English-speaking world knew what the Frankfurt School was. It was a footnote in books on intellectual history by liberal scholars like H. Stuart Hughes, not a presence to be reckoned with. *One-Dimensional Man* is still the only work of the Frankfurt School to have a significant political impact. The second amazing fact is that Marcuse's career as a philosopher begins with a publication in 1928 and continues till his death in 1979 with dozens of interesting articles and many books worthy of the attention of philosophers and intellectual historians. There is nothing wrong with this corpus of work that should cause it to be denigrated or ignored because Marcuse was briefly in the public eye. He did not go out to seek popularity. The people came to him.

There is an important difference between Marcuse and Adorno and Horkheimer that ought to interest us today. Marcuse alone understood the prefigurative significance of the New Left. He alone saw that new possibilities of opposition had emerged even as working class resistance to capitalism weakened. Surely, we need to understand this phenomenon, and Marcuse can contribute to that task, whereas I find less help in his Frankfurt School colleagues.

Index

A
Adorno, Theodor W., 7, 18, 40, 41, 323–326
Agency
human vs. technical, 24, 57, 141
AIDS patients, 266, 269, 275
Algorithms
and free will, 228, 232, 233
and organisms, 227, 233, 234
Alienation, 3, 9, 29, 169, 171, 180, 182, 220, 242, 256, 313
Anthropocentric (Anthropocene), 8, 11, 150, 152, 153
Anthropology (philosophical), 3, 4, 59
Aristotle, 19, 199, 200, 208
Articulation, 118, 119, 124, 125, 127, 128, 130, 131, 133, 135, 164, 172, 289
Artificial Intelligence
and free will, 232, 233

B
Bhaskar, Roy, 5
Biocentric, 151
Bolz, Norbert, 237
Bush, George, 150

C
Capitalism
communicative, 61, 164, 168, 169, 173, 176
market capitalism, 242, 251
postcapitalism, 261
Cavell, Stanley, 54
Christoff, Peter, 139
Civilization
model, 118, 131, 135
process of, 235
Civil society
global, 140, 154, 156
transnational, 140, 154
Class struggle, 9, 50, 164, 165, 167, 172, 182–184
Commodity (commodities)
fetishism, 29
form of, 29, 80, 267
Commoner, Barry, 145, 275, 290
Communication
digital, 8, 164, 167–169, 171, 173, 174, 176, 183

D.P. Arnold and A. Michel (eds.), *Critical Theory and the Thought of Andrew Feenberg*, https://doi.org/10.1007/978-3-319-57897-2

Communism, 144, 243, 255, 288, 319
Community, 7, 30, 31, 43, 249
Constructivism
radical, 167, 195
Consumption
green, 154, 155
Contemplation (contemplative attitude), 24, 27, 55
Controlled experiments, 101, 102
Crisis, 48, 50, 58, 62, 141, 152, 156, 194, 272, 312, 322
Critical realism, 299
Critical theory, 1, 2, 5–8, 10–12, 14, 18, 47, 48, 52, 53, 58–60, 71, 120, 129, 131, 142, 146, 152, 165, 166, 176, 181, 183, 194, 203, 217–219, 222, 224, 225, 229, 231, 242, 244–246, 248, 250, 254, 264, 265, 276–279, 286, 287, 303, 312, 321, 323–325
Crowell, Steven, 78
Cultural Transformation, 7

D

Dataism, 228, 234, 235
Dean, Jodi, 8
Deductivism, 101
Deep ecology, 145
Democracy
and technology, 156, 229, 313
green, 8, 139, 140, 145, 147, 150, 153, 155, 156
Dereification, 39, 322. *See also* reification
Determinism, 8, 9, 142, 173, 174, 183, 203, 272, 274, 275
Dewey, John, 54
Dialectic
as theory of social totality or culture, 58
Dialectic of Enlightenment (Horkheimer/Adorno), 40
Dickens, Charles, 143
Digital Age, 242, 247, 256
Disenchantment (Entzauberung), 218, 220
Dryzek, John, 8, 139, 141, 152
Dualism, 23, 24, 26, 33, 34, 37, 38, 41, 75, 180, 257
Dystopia, 6, 118, 119, 127, 133, 134, 195, 196, 198, 218–220, 222, 225, 230, 236, 247, 289, 290

E

Earthfirst!, 8, 154
Eckersley, Robin, 8, 152
Ecocentrism, 152
Economy, economic
green, 155, 156
system, 4, 6, 12, 34, 62, 146, 155, 156, 241
Ecocentrism, 152
Eco-primitivists, 141
Efficiency, 7, 9, 124, 126, 129, 134, 142–144, 196, 199, 220, 246, 313
Ehrlich, Paul, 142, 145
Empire, 163, 166–168, 172, 176, 319
Empiricism, 2, 21, 30, 176
Enframing, 194. *See also Ge-stell*
Enlightenment Project, 247, 255
Environmental
agency as non-human agency, 236
philosophy, 20, 25, 26, 42
Epistemology
social, 72, 77
Essence, concept of, 201, 223
Experience
and the life world, 11
and integrated networks, 234

F
Facebook, 14, 228
Feenberg, Andrew
  Alternative Modernity, 141, 207, 264, 270, 275
  Between Reason and Experience, 144, 194, 217, 221, 222
  Critical Theory of Technology, 1, 9, 141, 163, 165, 225, 263, 277
  Philosophy of Praxis, 2, 3, 18, 19, 48, 49, 53, 62, 63, 244
  Questioning Technology, 142, 194, 203, 204, 206, 223, 265, 272, 275
Foreclosure Theory, 8, 164, 165
Form
  of objectivity (*Gegenständlichkeitsform*), 4, 37, 49, 50, 58, 298, 307, 310
  social, 5, 29, 36, 49, 50, 321, 322
Foucault, Michel, 257, 267, 273, 320, 324, 325
Frankfurt School, 1, 2, 6, 7, 11, 12, 18, 40, 41, 48, 51, 72, 146, 147, 152, 164, 167, 218, 221, 241–247, 250, 251, 255, 256, 258, 263–268, 271, 279, 312, 320, 321, 323–326
Fraser, Steve, 257
Free Will, 228, 232
Fuchs, Christian, 8, 164

G
Ge-stell, 9–11, 13. *See also* Enframing, Heidegger
Giddens, Anthony, 149
Global Parliament, 8, 154, 156
GNP, 144, 147
Goodin, Robert, 148
Governance
  multidimensional, 279
Gramsci, Antonio, 19, 304, 319
Great Recession, 241, 251, 257
Greatest Happiness Index, 144
Green
  consumption, 154, 155
  democracy, 8, 139, 140, 145, 147, 150, 154–156
  political theory, 148, 152
  production, 155
Greenpeace, 8, 154

H
Habermas, Jürgen, 1, 60, 77, 152, 155, 195, 245, 246, 248, 252, 267, 321, 324, 325
Hackers, 182
Hajer, Martin, 139
Hansen, Mark B.N., 236
Harari, Yuval Noah (Homo Deus), 10, 219, 231
Harding, Sandra, 150
Hardt, Michael (and Antonio Negri), 8, 163
Harré, Rom, 99
Hegel, Georg Wilhelm Friedrich, 22, 33, 243, 245
Heidegger, Martin
  and critique of epistemology, 221
  Gelassenheit, 220
  The Question Concerning Technology, 9, 220
Heilbroner, Robert, 145
Honneth, Axel, 3, 47, 54–61, 267, 324
Horkheimer, Max, 18, 41, 152, 242, 323, 326
Hörl, Erich, 236
Human Development Index, 144
Humanism
  fate of, 229

I
Idealism, 2, 3, 23, 24, 38, 57, 248, 304
Ideal-type(s), 248, 258, 267
Ihde, Don, 140
Immanent Critique, 242, 245, 249, 253
In(dividual), concept of, 55
Information, 11, 14, 21, 154, 172, 173, 177, 228, 234, 237, 285
Instrumentalism, 8–10, 118, 129, 163, 165, 166, 179, 182–184, 196, 200, 278
Instrumentalization Theory (Feenberg), 10, 12, 13, 195, 203, 206, 208, 209, 299–302, 313
Internet
  communist, 173, 176
Isolatability, 5, 6

J
Japanese Modernity, 270
Jonas, Hans, 145, 152

K
Kant, Immanuel, 22, 33, 73, 243, 308, 310
Kellner, Douglas, 1, 12–14, 286
Kitcher, Philip, 150
Knowledge
  metacritique of, 49
Kuhn, Thomas, 143

L
Labor
  power, 84, 178, 306
Laclau, Ernesto (and Chantal Mouffe), 124
Lask, Emil, 74
Legitimation, 143, 149
Limits to Growth (Club of Rome), 142
Longino, Helen, 150
Lukács, Georg
  History and Class Consciousness, 4, 47, 304, 322

M
Marcuse, Herbert
  One Dimensional Man, 9, 150, 246
Market, 29, 31, 43, 61, 165, 272, 297, 302
Marx, Karl
  capital, 4, 26, 320
Marxism
  western, 8, 18, 163, 170, 217, 218
Materialism
  activist, 24, 28, 35
Mathematisation, 101, 105
Meadows, Donella, 142
Media technologies, 236
Mediation, 3, 39, 51, 62, 178
Minitel Videotext system, 266
Modernism
  strong ecological, 139
Modernity, 13, 48, 53, 171, 201, 217, 218, 221–223, 225, 227, 229, 231, 232, 235, 242, 253, 255, 264, 270, 273, 276, 280, 300
Multitude, 9, 165, 168, 177, 179, 183, 205

N
Natural Science, 33, 36, 37, 40, 291, 300, 305–307, 309, 311, 312
Nature
  reconciliation with, 51, 52
  reification of, 304, 310, 313

second, 34, 55, 57, 220
New Left, 19, 41, 52, 60, 244, 245, 275, 325, 326
National Science Foundation (NSF), 296

O
Ontic, Ontological, 37, 49, 52, 59, 72, 119, 125, 127, 134, 169, 194, 197, 202, 287, 289, 306, 311
Ontology
social, 4, 9, 72, 131, 199
Open and close Systems, 105

P
Phenomenology, 2, 9, 10, 73, 176, 194, 195, 200, 203, 321
Political economy, 5, 72, 73, 86, 251, 322
Political Theory
green, 148, 152
Popper, Karl Raimund, 12, 251
Positivism, 71, 252
Postmodernism, 2, 9, 12, 241, 242, 250, 257
Pragmatism, 241
Praxis, philosophy of, 18, 19, 48, 49, 51, 59, 62
Praxis (practice), 1–3, 19, 319, 322, 323
Proletariat, 12, 29, 50, 166, 257, 286, 322

R
Ralston, Holmes, 159
Rationalism
adminstrative, 149, 150, 291, 292
Rationality (rationalization)
instrumental, formal, calculative, 49, 298
technological, 7
Recognition
antecedent or original, 56
forgetfulness of, 53, 56, 59
theory of, 59
Reification, 3–6, 29, 320, 322, 323. *See also* Dereification
and the philosophy of the Subject, 53
as forgetfulness of recognition, 56
in the Philosophy of Praxis, 3, 297
Retroduction, 100
Revolution
and the realization of philosophy, 49
failure of, 51
Ricardo, David, 243

S
Schiller, Friedrich, 22
Science
and mathematics, 97
theories of, 105
Simondon, Gilbert, 96, 142
Singer, Peter, 151
Situational logic, 256
Social
engineering, 37
logics, 122, 124, 125, 127, 130–132, 134
ontology, 4, 9, 72, 199
rationality, 5, 7
Socialism
state socialism, 243
Substance metaphysics, 195, 205
Sustainability, 7, 124, 252
Synergistic totality, 142

T
Techne
  and *poiesis*, 222
Technical
  code, 6, 52
  design, 13, 145, 147, 203, 235
  devices, 13, 194, 195, 200, 203, 205, 219
  politics, 6, 53
Technique
  boundary of, 119, 121, 130, 135
Technology
  and democratization, 52, 118, 128, 145, 194, 276
  and value neutrality, 167, 179, 195, 203, 293
  converging (nano-bio-info-cogno), 219, 232
  in modernity, 225, 274
  technological revealing, 220–222, 225, 229, 235
Techno-Humanism, 228
Technophobia /Technophilia, 250
Technoscience, 242, 248, 250, 251, 253, 255, 258
Tragedy of the commons, 30
Transformation, structural, 7, 144, 154, 155
Trump, Donald, 150, 280, 324

U
Utopia
  utopian construct, 245

V
Value
  aesthetic and ehical, 13
  civilizational, 143
  exchange value vs use value, 29, 80, 82, 83, 172, 302
  value form theory, 72
Vico, Giambattista, 22
Vogel, Steven, 1–3, 44, 159, 303, 305, 306, 308, 311, 313, 316

W
Weber, Max, 217, 220, 269, 320, 321
Women, childbirth, 266
Working Class, 164, 170, 252, 323, 326

Z
Zoocentrism, 151, 153

The manufacturer's authorised representative in the EU is Springer Nature Customer Service Centre GmbH, Europaplatz 3, 69115 Heidelberg, Germany. If you have any concerns regarding our products, please contact ProductSafety@springernature.com

Printed and bound by CPI Group (UK) Ltd, Croydon, CR0 4YY
25/09/2026
02222890-0008